EMERGENCY POWER
FOR RADIO COMMUNICATIONS

by Michael Bryce, WB8VGE

Published by **ARRL** The national association for **AMATEUR RADIO**

Newington, CT 06111-1494
ARRLWeb: www.arrl.org

Production
Michelle Bloom, WB1ENT
Jodi Morin, KA1JPA
David Pingree, N1NAS

Cover
Sue Fagan

MW00851295

Copyright © 2005 by
The American Radio Relay League, Inc.

*Copyright secured under the Pan-American
Convention*

International Copyright secured

This work is Publication No. 315 of the Radio
Amateur's Library, published by the ARRL.
All rights reserved. No part of this work may
be reproduced in any form except by written
permission of the publisher. All rights of
translation are reserved.

Printed in the USA

Quedan reservados todos los derechos

ISBN: 0-87259-953-1

1st Edition
1st Printing

We strive to produce books without errors. Sometimes mistakes do occur, however. When we become aware of problems in our books (other than obvious typographical errors that should not cause our readers any problems) we post an Adobe Portable Document Format (PDF) file on *ARRLWeb*. If you think you have found an error, please check **www.arrl.org/notes** for corrections. If you don't find a correction there, please let us know, either using the Feedback Form in the back of this book or by sending an e-mail to **pubsfdbk@arrl.org**.

Contents

Foreword

Emergency Communications is an area that has become more important than ever. Many aspects have been discussed pertaining to the proper equipment to use; efficient antennas, the latest in transceivers and their installation, along with appropriate operating technique. This book covers, in detail, the foundation of any communications installation — the power source.

Without a source of electricity, there is no communication as we know it. We are dependent on the power grid, but with the advent of recent long-term outages due to weather and problems with the grid itself, we are increasingly aware of the possibility of long-term power outages. There have always been situations in which areas were beyond the reach of the commercial power grid, resulting in alternative electric power generation. Now that the outages have become more widespread, we need to be aware of alternative methods of generating our own electricity.

This book covers these alternative methods of power generation. The portable hand-held transceiver that is used with its internal battery pack still needs a source of power to recharge it, especially when the grid is unavailable. Keeping the lights on so that you don't have to communicate in the dark is discussed — there's more to emergency communications than just running the radio equipment!

Mike, WB8VGE, covers the various means of electric power generation, including advantages and disadvantages of different methods. There is no single "best" method of alternative power. Based on varied situations, this book will help identify the methods that will work best in your particular situation, perhaps taking advantage of possibilities already on hand.

In the more than 20 years that I've known Mike, he has always had the means to function without the power grid when needed. This book is a result of his knowledge and experience.

Steven W. Garwood, NØCZV
Telecommunications Systems Analyst II
State Of Ohio — Multi-Agency Radio Communications System (Marcs)

Preface

Katrina.

One word that is engraved in the memory of hundreds of thousands of people. Not just a word, but also a cry of despair.

In the final weeks of preparation for this book, the name Katrina flooded the TV, radio and waking hours for millions of people. This was no simple power outage; it was truly a catastrophe of Biblical proportions. It also became one of Amateur Radio's brightest moments in providing life saving service and communications to countless people.

In the early days after Katrina, hams often were the only means of communication left as rescuers plucked people out of floods, off rooftops, and from cars. But the power did not come back on. Hams then turned to assist the relief agencies ... but still the power did not come back. Hospital evacuations, looting, distribution of medicines, huge and horrible "shelters of last resort" filled the news. Still, no power. Through it all, the hams were there, doing what hams do, somehow making the messages get through.

Katrina was also a great teacher if we will listen to her lessons. We learned that it will not be the small problems that call most urgently for Amateur Radio's aid, but the large ones. That aid will probably come in from a distance because the local hams are now victims themselves. These incoming volunteers will need to bring electric power with them. Not power for one day or even three, but possibly for a week or more.

Katrina taught us that we need to be self-reliant. This book, in a very timely way, helps teach us how to achieve part of that goal. We are, after all, the Amateur Radio *Service*. In the dark days of September 2005, we showed that the last word in our name, often forgotten, can be the most important word of all. This book can show you how.

David Sumner, K1ZZ
Executive Vice President, ARRL
Newington, Connecticut

Acknowledgements

The fact that you're holding this book in your hands says a lot about you. You have adventure in your blood and an independent streak. Self-sufficiency and the ability to overcome obstacles on your own is something in which you take great pride. You don't like being in the dark.

Putting together a book that is both informal yet provides an in-depth look at various power sources might seem to be an easy task. It was not.

There is more to it than just sitting down in front of a computer and typing away. There are endless hours of snapping photographs, turning wrenches, adjusting controllers, checking the water in the batteries and throwing tools against the garage wall. But it was a lot of fun!

It is the *people* that bring a project like this to fruition, however. It is time that those closest to me are acknowledged for their help.

My wife Donna has never complained about sitting in the dark while I reset breakers or changed out batteries. She has repaired countless holes in my clothes cause by battery acid. From eating somewhat hot food in front of the wood burner to keeping the lights off when the batteries are low, she has always been there. Donna, I would marry you all over again!

On Easter Sunday of 2005, I lost my son Christopher. He was 22 years old. He was the source of a lot of my personal energy. His dad will miss him beyond words.

Steve, NØCZV, would always listen to my extreme off-the-wall methods of keeping the lights on. Always one to try new things, Steve has slowly become more and more energy dependent. While you won't find him working the CW station at Field Day, he will be manning the 60 Hz station.

A tip of the hat to Republic Engineered Products.

For all the hams that set up stations and got on the air with emergency power before, during and after Hurricanes Katrina and Rita in 2005, thank you.

Finally I would like to extend my deepest gratitude to my editor, Larry Wolfgang, WR1B. It was his skill that took countless emails, digital photographs and hand scribbled notes, and assembled them into this book.

Mike Bryce, WB8VGE
North Lawrence, Ohio

About the Author

Mike Bryce, WB8VGE, of North Lawrence Ohio has been on the air since 1975. He currently holds an Amateur Extra class license.

An avid QRPer, Mike has been inducted into the QRP hall of fame. He edited the QRP column in *73 Amateur Radio Today* magazine for 12 years He also has authored numerous articles in *QST, CQ, HomePower, 73 Amateur Radio Today* magazine , *Nuts and Volts, CTM,* and *The QRP Quarterly*. He was past president of the Massillion (Ohio) Amateur Radio club.

Mike is currently employed by Republic Engineered Products.

Mike has always been fascinated with energy, particularly solar energy. His solar-electric setup can easily run his entire home. If you

hear WB8VGE on the air, you can bet the farm his signals came from energy captured from the sun.

When Mike is not checking battery water, he enjoys restoring Heathkit and Drake radio equipment. One cannot live by tubes alone, so Mike has acquired a nice collection of Ten-Tec radios as well.

While working on vintage radios can be a lot of fun, sometimes it is a great stress reliever to heat a bolt until it is red-hot and then twist it off. To that end, Mike enjoys working on his Jeep vehicles. He currently own two YJ's a TJ and a KJ.

If Mike is not in the shack or working on the Jeeps, then he can be found in his greenhouse. For some unknown reason, Mike as a passion for sunflowers!

Chapter 1

Keeping the Signals on the Air

We have all become addicted to electrical power. We have become power junkies. We need electricity to pump our water, keep our food fresh and process our waste. Many youth today have no idea that popcorn can be made in a skillet on top of the stove or over an open campfire. It's all done via the microwave now! You come home, flip a switch, turn on the TV and fire up your HF gear for a chat on 75 meters. The power is always there, just waiting for you to command it. As with any addiction, there's a price to be paid. When the lights go out everything grinds to a halt. And things do go wrong.

There are very few things in life that are certain. Of course we all know that you can't avoid death or taxes, but emergencies have a knack of occurring when you least expect them and usually when you are least prepared. They often catch you "with your pants down," just like on a hot August day in 2003.

THE GREAT 2003 BLACKOUT

So, where were you on August 14, 2003? Well if you were living in the northeastern part of the country, you found yourself sitting in the dark along with 50 million other people. According to the studies and inquiries done by independent groups and the affected utilities, tree branches touched several of the high-voltage lines in Ohio. FirstEnergy of Akron had allowed trees and other vegeta-

tion to grow too close to the line. At the time in August, heavy electrical loads on the line caused it to sag, hitting the tree limbs.

A Failure In The Making

As utility after utility went off line, the entire electrical power grid failed in a cascade that took less than one minute. With only a few exceptions of small islands of power, most of the northeastern part of the United States was without power.

Not only had the power grid itself gone down, power generation plants tripped off line as well. There were safety systems in place that took the plants off of the grid to prevent damage to generators and switching gear. But as more and more generation plants went off line, the power grid became even more loaded. As the load increased, the last few plants dropped off line because of the overload on the grid. Once the process started, it was almost impossible to stop.

To make matters worse, you just don't flip a switch and start the power flowing again. The entire grid must be brought back up a little at a time and in a precise order. It took almost three days to get all the power back on. To this day, FirstEnergy denies the problem started with their system. The bottom line, however, was that a lot of people were sitting in the dark. We had an emergency on our hands that affected a lot of people.

August 14, 2003 Blackout Timeline

3:05:41 PM	FirstEnergy 345 kilovolt line fails because of a tree
3:32:03	FirstEnergy 345 kilovolt line fails because of a tree
3:41:33	FirstEnergy 345 kilovolt line fails because of a tree
4:05:57	FirstEnergy 345 kilovolt line fails, triggering the blackout
4:08:59 to 4:10:39	Power plants start shutting down: transmission lines disconnect in northern Ohio and Michigan.
4:10:38	A FirstEnergy transmission line in Northeast Ohio trips, severing the path into northern Ohio from Pennsylvania. Because of that, power reverses direction from Pennsylvania to New York, Ontario and Michigan.
4:10:39 to 4:10:46	In Ohio and Michigan, conditions keep degrading. More power plants, including Perry nuclear plant begin shutting down. Power lines disconnect in New Jersey and Ontario. Western Pennsylvania separates from New York.
4:10:46 to 4:10:49	Power line from New York to New England disconnects. New York transmission splits.
4:10:50	Ontario blacks out after disconnecting from New York
4:11:22	Southwest Connecticut separates from New York city.
4:13:00	The cascading blackout is essentially complete, with more than 100 power plants shut down and as many as 50 million people without electricity.

Source: U.S.-Canada power Outage task force. All times are PM Eastern Daylight time.

Crisis In The Making

In a short time, the backup batteries of the landline-based telephone system were drained. The telephones stopped working. Cell phones could not operate when there was no power to the cell sites. Local police and public service communication systems were all jammed. Hospitals were running on emergency generators while others were trying in vain to get emergency power established. We were in the middle of a national power emergency.

Before long, a lot of those 50 million people were out of water, out of ice and worried about frozen food rotting in their freezers. They could not contact their loved ones, because most forms of communications were either down or jammed beyond belief. This was a time for ham radio operators to do what they do best: set up emergency communications in the field, operated by emergency backup power sources. And that's exactly what we did!

Just What Defines An Emergency?

Someone once asked Mel Brooks what defines comedy. He said it was a matter of reference. For example, if *you* fall down a flight of stairs that's funny! If *I* get a paper cut, that's a tragedy.

There's no question that on August 14, 2003 we had an emergency that affected a huge chunk of the country. But, if you lived in Washington state, you had nothing to worry about. The power outage emergency did not affect you in any obvious way.

I found out how true that was on August 14, 2003. Since I had power, thanks to my solar electric systems, I checked into several HF nets. I mentioned that the city of Cleveland Ohio was running out of water. The net control station thanked me and then went to his next station. Looking back, there was no need for me to waste power (and the net's time) by checking into a net to tell them about the problem in Cleveland. He was in no position to do anything about it.

That's the problem with any emergency. It's usually very local. When hurricane Andrew went though Homestead Florida, that was an emergency on the highest scale for those living in and near Homestead. But for me in Ohio, I was unaffected by it all. Those hams in and around Homestead were pressed into service trying to provide communications using emergency power of all sorts.

While it is true that during and after Andrew went through Florida, some communication was done on the HF bands, the majority of work was done using our VHF and UHF bands. Again, this is due to the nature of the local emergency.

OPERATING IN LOCAL EMERGENCIES

Nearly all emergencies are of the local nature. We normally use VHF and UHF bands to handle the communications in such cases. That's good from an energy point of view. A 10 to 15-W VHF transceiver will require less power to operate than a standard 100-W HF station. That's no surprise to anyone, but with VHF transceivers now pushing the 75-W output level, we need to consider these high power units in our energy budget.

One benefit of using VHF and UHF is the ability to use directional antennas to increase range. Also, there may be

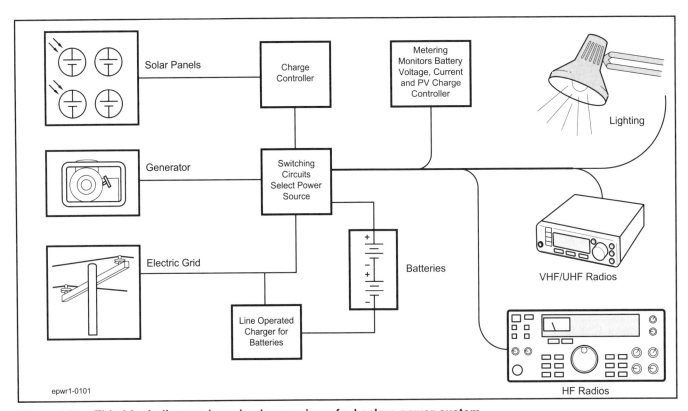

epwr1-0101

Figure 1.1 — This block diagram is a simple overview of a backup power system.

some repeaters that will function during a local emergency. Don't count on this for a backup system, however. You never know what repeaters, if any, may be working during a power outage.

Putting a signal out into the air requires power. It doesn't matter if you have the latest high tech DSP radio equipment, it is just refined silicon and plastic without the power to make it work.

NO SINGLE SOURCE IS A PERFECT SOLUTION

When it comes to supplying power when the grid is down, one thing is certain. You never know when it might happen. To make things even more interesting, you really don't have any idea how long it will be before commercial power is restored. In the case of ice storms, it could be weeks before the lights blink back on. When the August 14 power outage hit my neighborhood, it took over 14 hours for the utility to restore my power. Lucky for me, I had solar power to operate my station and most of the house. What works for me in my area may not be suitable for you, though. Solar power requires access to the sun's rays. If you live in an apartment, solar power may not be the best choice for emergency power. Of course, if you have access to the roof to set your panels, then you're good to go. **Figure 1-1** is a block diagram of the backup power system at my house.

Let's take a quick look at some of the various methods of providing emergency power to operate your ham station. I'll start off with the source of just about all energy on the planet: solar power.

Solar Power For Emergency Power Backup

I have to admit it. I am biased towards solar power. My ham radio station has been running from the sun's power since 1978. Solar energy provides all the power I need in any power emergency. See **Figure 1-2**.

My solar arrays convert the sun's energy into direct-current electricity by using solar panels. These photovoltaic (PV) panels instantly convert the sun's rays into direct current while producing zero emissions of any kind. They con-

sume nothing in the process. With no moving parts, they produce power silently while sitting in the sunlight. This seems like a perfect source of power, and yet solar panels do have some drawbacks. First, you must have sunlight falling on the cells. No glow, no go! Second, they are expensive when compared to other sources of power. Third, you need a lot of them to produce useable power.

Of course you have to store that electricity so you have some available after dark or on cloudy days. **Figure 1-3** shows a bank of batteries ready to be wired for a backup storage system. These batteries will be wired to provide a 24-V system with a capacity of 440 Ah.

Solar arrays are available in many sizes. They aren't all as big as the ones shown in Figure 1-2. For example, the small 1-W panel shown in **Figure 1-4** could be used for a

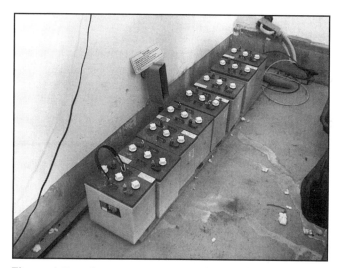

Figure 1-3 — Any emergency power system needs batteries. Here, eight 6-V batteries will be wired in series-parallel for a combined power of 24 V at 440 Ah.

Figure 1-2 — These large solar panels charge the batteries that keep WB8VGE on the air without commercial power.

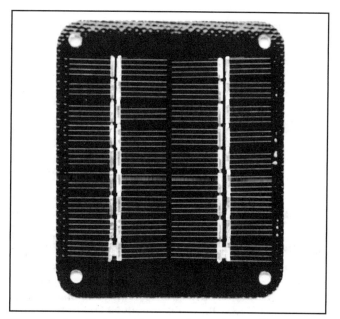

Figure 1-4 — Not all solar panels are huge. This panel produces a whopping one watt! It's ideal for backpacking or hiking.

Figure 1-5 — Solar power is providing the electricity for my remote APRS weather station. The solar panel is a Solarex VLX-20. This panel provides all the power necessary to run an APRS packet station 24 hours per day, 7 days per week. The controls, battery and radio gear are mounted in the box on the pole.

Figure 1-6 — This is about as simple as it gets: a solar panel, battery and a QRP transceiver. The solar panels is a Solarex MSX-10L, which produces 10 W under standard test conditions. The transceiver is a 5-W, CW-only rig. The charge controller, key and a few other station items are not shown.

variety of low-power applications. Small solar arrays are useful for powering equipment in locations where it is not convenient to run power lines. **Figure 1-5** shows my remote wireless APRS weather station. The small solar array produces enough power to keep that station operating around the clock. **Figure 1-6** shows a small panel with a battery and QRP radio that makes a great station to take camping.

When you really like to use the sun's energy, you can find all sorts of ways to do it. **Figure 1-7** is a photo of my two electric lawn tractors. No loud gasoline engines and no high gas prices to worry about with these tractors. I just recharge the batteries with the solar array, and they are ready to go the next time I have to mow the lawn!

Generators

One thing that makes solar power useable in locations at night and under limited hours of sunlight is a backup generator.

Generators produce electricity by moving a magnetic field within the generator's windings. To spin the magnetic field an internal combustion engine burning either gasoline, natural gas, propane or diesel fuel is normally used. (It is possible to spin a generator using wind or water power, but those sources are seldom used for an emergency backup system.) Some generators will operate on natural gas or propane and gasoline at the flip of a lever. These so-called tri-fuel generators are at the top end, and thus very expensive. The bottom line is that generators require some sort of fossil fuel to produce electricity.

That's the biggest problem with generators. They need fuel to operate. It's hard to store fuel safely and you never know how much to store. **Figure 1-8** shows a classic China Diesel generator that produces 10 kW of power.

They also produce exhaust gases that must be vented. Never, ever operate a generator in an enclosed space. They produce carbon monoxide, and that will kill you. So you see, generators are not the perfect source of emergency power either.

Figure 1-7 — While not exactly used during an emergency, my General Electric ElecTrac tractors can be charged by the solar array in the background. Each tractor holds six 6-V batteries rated at 220 Ah each.

The Amherst Wisconsin Energy Show

I had a chance to set up my solar electric systems and control units at the Amherst Wisconsin energy show one year. It's a great place to stop and see all the stuff that is available for generating electricity off of the grid.

There were two items that dripped "cool." One was a Sterling engine that was making power from burning wood pellets. The ones that really took in everyone at the fair, though, were the steam engines!

Photo A shows two versions of the same basic engine. On the right side of the photo, the twin "V" sports a clutch and a crankshaft to couple the two cylinders. In both engines, the steam is used to push the cylinder up and then back down. The steam is then exhausted out of a port. This is called an open system. The exhausted steam in not reused. To provide enough steam to drive the engines, it took a *lot* of heat and a *lot* of water!

In a closed system, the steam is trapped, cooled to condense it back into water, and then sent back to the boiler. It still takes a lot of heat to produce the steam, but you don't need a constant supply of fresh water.

The steam engine shown in **Photo B** would drive a 5-kW power take off generator. The fly wheel weighs over 500 lbs. In the background, you can see the boiler. Just in front of the boiler's stack, you can see the mechanical speed governor. The pipe that is across the photo is the water supply line going to the boiler.

Photo C shows a steam engine driving a car alternator. This system is able to produce over 60 A of charging current. You can't see them in the photo, but the system is charging four deep-cycle golf-cart batteries.

These working examples of making electricity from steam are very "cool" but they are also incredibly inefficient. Steam engines are not something you just throw together. You can be killed or seriously burned by an exploding boiler. Let the big boys boil the water at the nuclear plant. There are many other practical ways to make electricity, and they are much safer for a home operation.

Photo B — This monster will drive a 5-kW power take off (PTO) generator such as might be used with a farm tractor. Again, an this is an open system, which requires a constant water supply.

Photo A — Two cylinders or one? The steam engine on the left is a single cylinder, which produces 20 hp. The twin V opposing cylinders shown on the right produce about 40 hp. Both of these engines are open systems. In other words, after the steam is used, it is discarded and not recycled back into the boiler.

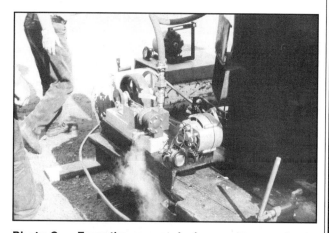

Photo C — From the newest design nuclear reactor to coal fired power plant, the number one way of producing electricity is by heating water to make steam. This steam engine is driving a car alternator. At the time this photo was taken, the combination of the two had charged up a set of four deep cycle golf cart batteries. The charging current was in excess of 60 A!

Figure 1-8 — Here's a classic. This China diesel engine and generator produces 10 kW. It can supply either 120 or 240 V ac.

Figure 1-9 — This little Windseeker can provide a surprising amount of power, if supplied with enough wind! It is shown here on display at the Amherst Wisconsin Energy Show. In actual use, the Windseeker unit would be mounted on a tower high enough to catch the prevailing wind.

Figure 1-10 — An emergency back up system need not be complex. This simple system was designed for both wind and solar generation. Eight Exide 6-V deep cycle batteries are configured for a 12-V system. Total battery capacity is just under 1000 Ah.

Micro Hydro Power Systems

The power of falling water provides this country with millions of megawatts of electricity each year. On a much smaller scale, however, we have what is known as "micro hydro" systems. For most of us, the use of a micro hydro power plant is just a daydream. We don't have the access to the necessary water, or the land to hold the equipment.

To make use of the power of water, you need both volume and flow. You also need to know the fall of the water source. Knowing all three of these factors will determine if you can use a micro hydro system. There are other factors that need to be examined too. Most small creeks only flow in the spring and fall. Then these creeks dry up or produce so little flow you can't produce power from them. Of course, you can't just dam up a creek. The local county and state

departments of natural resources would be pounding on your door if you decide to stop the flow of a waterway.

So its time to get real. Micro hydro just won't be a valuable source of emergency power except for a small fraction of ham radio operators.

Watts Blowing in the Wind?

Wind power on the other hand, is a possibility for a lot of us, provided we have the land to hold the wind-turbine tower and the required wind. In most cases you can't produce any usable power with wind speeds under 12 mph.

Wind power is romantic! It's hard not to look up and watch the blades spinning in the wind. Doesn't it give you a case of the warm fuzzies?

Will wind power produce the required power for your shack? Yes, it can, provided you have the wind and location. Wind generators run from the small AIR 400 turbine to larger 10 kW units. **Figure 1-9** shows a small Windseeker on display at the Amherst, Wisconsin Energy Show. Mounted high on a tower, this unit will generate a surprising amount of power.

You can find some of the older Jacobs wind generators

still working in the plains states. They are 32-V systems, so watch what you are getting into. You will also need a rather high tower to hold the generator. Make sure you enjoy climbing the 90-plus feet to do the necessary maintenance twice a year!

The Best of Both Worlds

Since you can see that no one source of power is the perfect pick, we can combine two or more in a hybrid backup system. One popular hybrid combines the use of PV panels and a wind turbine. It works like this. Normally during the hot dry days of summer we have lots of sun, but very little wind. But during the winter months and spring — when the sun's rays are hiding behind clouds — we have plenty of wind to produce power.

A backup generator, powered by a fossil fuel can replace the wind turbine. In both cases a hybrid system uses a large battery bank to hold the power for a rainy day. See **Figure 1-10**.

Later in this book, we will consider some of the more popular methods of generating electricity. It's definitely not a case of one size fits all. Putting that signal on the air sometimes requires more than one source of back up power.

In the pages ahead, I'll look at the various forms of generating electrical power when the power grid is down. I just touched on some of the most popular ones. I'll go into more detail on solar power and wind power. We'll take a closer look at generators, too.

We will discuss various sources of power, as well as all the little things that make up a dependable and safe emergency power system. We call these the "balance of system" components. At first most of us think of batteries and racks of solar panels when we talk about emergency power. Something as simple as a missing fuse can bring things to a quick halt, however.

It's like the family that went out for a bit of four wheeling and got hung up on some rocks. The vehicle they were driving had an electric winch that would have easily pulled them out of trouble. The family was stranded for over a day until someone else came along to help them. What was the problem that almost killed this family? Well it was a com-bination of several things. First, you never do anything like that alone. Second you tell others where you will be going, when you plan on getting there and when you plan on returning.

What was that "one little thing" that would have saved the day, though? They forgot the winch controller! Hopefully that won't happen to you when you plan your own emergency power system.

Of course you don't have to wait for something to go wrong with the power grid either. We will take a look at portable operation as well as complete stand-alone systems.

SAFETY

There is one subject you will hear me talk a lot about: safety. It's so very easy to become complacent when dealing with *just* 12 V dc. This is such an important subject, I've devoted an entire chapter on installing your emergency power system as safely as you can build one. I'll say it now and I'll say it several more times thoughout this book. "You can burn your house down just as quickly with 12 V as with 120 V."

Putting it all Together

While making electricity is number one on our list for emergency power, I'll also show you several examples of distribution of that power. It's not as simple as just twisting two wires together. Low voltage dc has a mind of its own.

When we need lots of power, low-voltage dc is just not up to the task. In cases like this, we need to change the dc over to something more usable like sine-wave ac. So there are methods that are quite simple, others verge on the overly complex. We will look at them all.

There will be plenty of "rules of thumb" to go by. I've discovered during the last 26 years that there are many ways of running equipment from solar and wind power. Some work, some work really great and others just simply stink.

It's going to be a wild ride! Sit back and enjoy. If you put to use some of the ideas I will be showing you thoughout these pages, you'll be on the air the next time the trees need trimming in Ohio — or wherever the next major power blackout begins!

Chapter 2

Hey, I am in the Dark:

Keeping the Lights on in the Ham Shack with Emergency Power

Somewhere on this planet, there is a law carved into stone that says when the power goes out, it will be at night. At least that always seems to be the case with me.

I remember as a kid living on the farm, the storms would roll though the corn fields on those hot May afternoons. The sky would darken from a light blue to an ever deepening blue. You could sit on the porch and watch the sky turn black. It was as though the finger of God was painting the sky, and the sky would appear to take on a life of its own. The lighting would be too far away to see but you could hear the rumbles of thunder coming from the west.

The power lines to our house were old and lacking any insulation. When the sky turned black, you could quite easily see the white power lines against the deepening sky. It was an awesome sight that I can recall as clearly today as when I was a small child.

It was during one of these storms that I fell down the cellar steps and broke an arm. With nothing to light the way, I fell down all 12 steps. From that day on, I've been especially aware of keeping the lights on when the power goes off.

Humans just don't like to be left in the dark. We are drawn toward a light in the darkness like a moth to a flame. We seek it out. It gives us comfort and a sense of security. Lighting keeps the bad guys at bay and the critters away. No one wants to be in the dark!

At first, lighting would seem to be rather unimportant when talking about keeping your station on the air. Having effective, efficient lighting, however, can make a huge difference in how you manage your station during a power outage.

It's one thing to operate your gear during the day on battery power but it becomes an entirely different matter once the sun sets. It's just plain hard to do anything when you can't see what you're trying to do.

THE OLD STANDBY: THE FLASHLIGHT

The first thing most of us do is grab a flashlight. That's cool, but they're really hard to use when trying to write on a tablet or trying to read a frequency on a radio display. The bright light will wash out the display much like a camera flash.

Today we have flashlights that outperform units just a few years old. Three important technologies have changed the personal lighting business.

First, the lamps themselves have changed. Light is created in an incandescent bulb by burning a filament in a vacuum. Current from a battery is passed through a filament, which glows white, making some light and a lot of heat. That has hardly improved since the days of Edison. Today we still burn a filament, but instead of in a vacuum, it is often burning in either a xenon or halogen atmosphere. The results are simply amazing! The popular Maglite (**www.maglite.com**) 2-D flashlight had an increase of 59% in candlepower using the xenon lamps. The halogen bulbs used in the rechargeable Maglites produced an average of 33% more light. **Figure 2-1** shows three Maglites. **Figure 2-2** shows a Pelican Pocket Sabre flashlight. This one uses two C cells and produces a very bright light because it uses a xenon lamp.

The second technology nudge came from battery chemistry. Although

Figure 2-1 — Don't overlook the good 'ol flashlight for emergency lighting. These Maglite flashlights come in all sorts of sizes and colors. I prefer the C-cell size. In many cases, you will be able to get replacement batteries from the corner gas station after everyone else has picked over all the D cells.

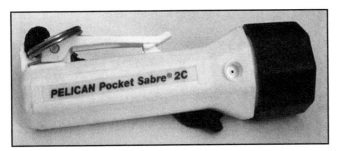

Figure 2-2 — A Pelican high tech flashlight provides blinding light with just two C cells.

most of us use primary batteries in our flashlights, higher capacity nickel cadmium (NiCd) and nickel metal hydride (NiMH) batteries are being put to use. Besides being rechargeable, the longer run times of the new NiMH batteries are now exceeding those of alkaline batteries.

Alkaline batteries are constantly being improved, too. The newer chemistry allows for some really serious run time. Every ham should have a fresh supply of alkaline batteries available.

Lithium batteries are making lights even brighter. Combine a lithium battery and bright-burning xenon bulb technologies and you end up with a weaponsgrade flashlight. An excellent example is the M6 Guardian from SureFire (**www.surefire.com**). Powered by six lithium batteries, the high output lamp provides 500 lumens of brightness. In layman's terms, that's about 30 times brighter than a typical two-D flashlight.

Figure 2-3 — Using only one "AAA" battery this LED flashlight produces such a powerful beam, you can't look directly into it.

Like everything else on the cutting edge of technology, lithium batteries are much more expensive than alkaline batteries. They may also be a lot harder to purchase in an emergency. Just about every small-town post office/general store/gas station/ice stand has some "flashlight batteries" hanging from the pegboard. Odds are about zilch that a replacement lithium battery would be in stock at a place like this.

The lastest technology advance has been made in light-emitting diodes (LEDs). While not producing the retina-bleeding beam of a xenon bulb, they produce a very bright light. What's more, usually the LEDs are in clusters, forming a powerful beam of light. **Figure 2-3** shows a small flashlight that uses a single AAA battery and super bright LEDs to produce a very bright beam. Mate a lithium battery with 1-watt ultra-white LEDs and you have a formidable light source that fits in the palm of your hand.

One great feature of an LED light is the long run time.

It's not at all unusual for run times that can be measured in *days*, not *hours*. In spite of these advances, a flashlight, either powered with a xenon bulb or a cluster of LEDs, really is not the best light source in the shack; they are too bright and too directional. There are other ways to brighten the night. Some of these work best for long-term use while others will serve us for just a short time.

CHEMICAL LIGHT STICKS

Just about everyone has seen these chemlights, as they are sometimes called, at Halloween time. They are quite simple to use and produce a nice bright light from half an hour to over twelve hours at a time. See **Figure 2-4**.

The chemical light sticks are really quite simple. Inside the plastic tube is a small glass vial. Bending the tube breaks the glass, allowing the chemical inside the glass vial to mix with another chemical in the plastic tube. When the two chemicals combine, light is produced. The resulting light is very much like that produced by a firefly, with no heat being produced. Because the light sticks don't produce heat, or have open flames, they are excellent for use when refueling generators.

The color of the light produced depends on the two chemicals used. Almost any color of light can be produced, including white. The chemicals that produce a brighter light won't produce useable light for as long. The ultra-bright white sticks are good for about half an hour. The yellow and amber sticks produce light for up to 12 hours for the same size light stick.

While the yellow and amber colors produce the longer running light, their monochromatic light can make reading color text or photographs a bit difficult. Keep this in mind when choosing a light stick to use with a radio that has a digital readout. You may need to choose a color that will allow you to read the display.

To get the most out of your light stick, cut a Styrofoam coffee cup length-wise to make a reflector. You can purchase a special holder for the light sticks, but they're expensive, and the coffee cup works great. Two or more of these light strips will provide all the light you need to operate your equipment and keep up with the paper work.

Light sticks do have a disadvantage or two. Even if not used, they have a shelf life of a year or so. After that time, they won't produce light to the specifications on the package. I rotate my stock of light sticks every year. I just give the old ones to the neighborhood kids at Halloween.

Like any chemical machine, they don't like to be cold.

Figure 2-4 — A chemical "glow stick" great for those times you need to check on the gas level on the generator. There's no flame, no heat and no sparks.

If you're planning on setting up in the cold, light sticks may not provide the necessary light.

While light sticks can provide useable light, you may need more lighting. Multiple stations will require more lighting than will one guy in a pup tent.

FLUORESCENT LIGHTING

The RV industry created a need for low-voltage lighting. Years ago, nearly all lighting in an RV consisted of a brake-light bulb in a plastic housing. That worked, but the light produced had a yellow glow about it. They also consumed a lot of juice from the 12-V house battery for the amount of light they produced. There had to be a better way of making light from a low-voltage source. The low-voltage fluorescent light was the way to go.

Operating on 12 V, the newest line of fluorescent lighting fixtures are not only attractive but can produce up to 60 W in a single fixture. Installed in the ceiling, low-voltage fluorescent lighting can light up an entire room.

Most low-voltage fluorescent light fixtures run about 15 W to 32 W. That's a draw of 2.5 A from a nominal 12 V system. The higher the wattage, the more current from the power source.

Lighting a fluorescent tube is not difficult. You simply apply a high voltage to both ends of the tube, and it glows. It's just that simple. Now, to produce the required high voltage from a low-voltage source such as our emergency battery system, we need a dc to dc transformer. Of course there's no such thing, but we can run an oscillator that switches the dc on and off at a very high rate. Then by using a transformer to develop the high voltage, we can light up the fluorescent tube. It sounds easy, and it is. Portable fluorescent lanterns are also popular with campers. **Figure 2-5** shows a Heathkit lantern I have used for years. **Figure 2-6** is a picture of a newer-model Coleman flourescent lantern, along with a spotlight I use around the shack.

There is a down side to all this simple technology, though. The oscillator produces lots of square waves and square waves are full of harmonic energy. So, as soon as we turn on our 12-V fluorescent light fixture, we also turn on a rather strong

Figure 2-6 — This photo shows a Stream light box and a Coleman fluorescent lantern. I use one for spot lighting and the other for room lighting.

Figure 2-5 — Here is a classic Heathkit fluorescent lantern.

transmitter as well. The oscillator and transformer comprise what is known as a ballast or fluorescent tube driver.

So, herein lies the rub when using low voltage fluorescent lighting. Some — not all, but some — fixtures produce so much RFI that you can't run the light and the radio at the same time. Sometimes you can filter out the noise if the noise is coming from the supply wires. In **Figure 2-7** you will see a simple brute force filter for the 12-V supply leads. Nothing is critical and you can use whatever you have laying around in the junk box. In general, more inductance is better for L1 and L2. Try 47 turns of number 18 enamel wire wound on a coil form of about ½ inch. That will result in an inductor of about 7 µH.

I have also installed several large computer-grade electrolytic capacitors across the light supply line in an attempt to decouple the light fixture from the power source. I've used combinations of computer grade capacitors totaling up to 1 F. Make sure you use capacitors with a voltage rating of at least 20 V dc. Also, with one Farad of capacitance, bleeder resistors would be an excellent idea. A 10-kΩ, ½-watt resistor would draw about 2 mA at 20 V dc. Too

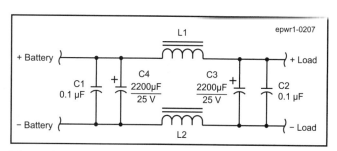

Figure 2-7 — You can use this "brute force filter" to filter out fluorescent lighting noise that appears on the power supply lines. See the text for a suggestion about coil construction.

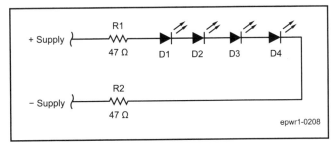

Figure 2-8 — This schematic diagram shows how you can connect three or four LEDs in series across your 12 V supply line. The 47 Ω resistors are a good starting point for the value of current-limiting resistors.

small of a bleeder resistor would add an extra phantom load to your battery bank. No matter what, you should always use a bleeder resistor across the capacitor bank. Even though the voltage is low, a one Farad capacitor can store a huge amount of energy. There's more than enough to burn the tip off a screwdriver!

If you want to try this, take a trip to the local Best Buy or Circuit City store. Another good place would be a high end automobile audio store. These stores will usually have 1 F capacitors rated for low voltage 12 V dc systems. They are used in car audio systems. They're not cheap, about $40 each, but that may be cheaper than stringing together a lot of odd shaped and mismatched capacitors from a hamfest.

Even with the large electrolytic capacitor bank, you can sometimes still hear the RFI from the light. Try using both the inline filter and the capacitor bank. The idea is to keep any harmonic energy from running down the light power supply lines into the power supply. If the noise is being radiated from the fluorescent tube itself, however, no amount of filtering in the supply line will fix the problem.

I had one 45 W lighting fixture that was so noisy there was nothing I could do short of turning it off to get rid of the RFI. On the HF bands the light produced well over S-9 signals across the bands.

If you are mainly using the VHF

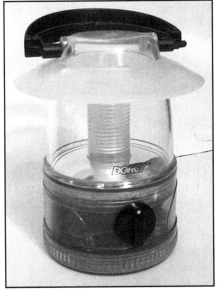

Figure 2-9 — I converted this Dorcy lantern from a regular light to a super bright white LED. While it is not as bright as the light that was in the lantern, the run time went from two hours to over a week with the same four AA batteries.

and UHF bands, you may be able to use a fluorescent lighting fixture that is producing RFI on the HF bands. FM rigs aren't normally bothered by the electrical noise produced by the driver circuit in the fluorescent light.

No matter what kind of fluorescent lighting fixture you have in your shack, it is best to turn it on and then try each and every radio and frequency you plan on operating should the need arise.

LED LIGHTING IN THE SHACK

We now end up full circle back to LED technology. With the new ultra bright LEDs, we can light up the entire shack and do so with a minimum amount of power. Best of all, the new ultra bright LEDs now come in ultra white. You get both bright lighting and full color retention. While there are several companies that market LED lighting, we can make our own fairly inexpensively.

Most white light LEDs require between 3 and 4 V at 20 to 30 mA. By putting three or four LEDs in series, we can make a rather bright light for very little time and money. **Figure 2-8**, shows a simple circuit for using LEDs in a string. Notice the current limiting resistors. The values shown are a good starting point. I used surplus LEDs and the current draw from any given string of three vary from lot to lot. The idea is to pick a resistor value that will hold the current to between 20 and 30 mA. Try a 47 Ω resistor to begin with. The resistor will get rather hot, so I use Xicon 5% metal oxide 2 W power resistors from Mouser Electronics (**www.mouser.com**). The part number is 283-47. **Figure 2-9** shows a lantern that I converted from an incandescent bulb to a super bright LED.

To find the correct resistor value, you can use a variable power supply set for the desired voltage at which you plan to run your emergency power system. If you plan to use batteries most of the time, set the supply for 12.5 V. Otherwise use the standard of 13.8 V. Then adjust the values of the resistors to match the LED current needs. The higher the resistance, the lower the circuit current will be.

We can take the project one more

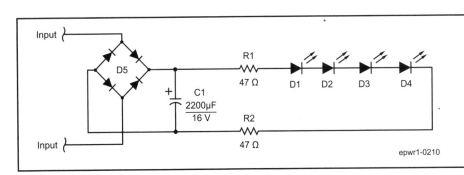

Figure 2-10 — This schematic diagram shows how you can use a diode bridge rectifier and capacitor to operate your string of ultra-bright LEDs from a low-voltage ac supply. For operation from a 12-V dc source the bridge rectifier eliminates the need to pay attention to the input voltage polarity.

step. By adding a diode bridge and a little bit of capacitance as shown in **Figure 2-10**, the circuit can operate from a low-voltage ac source. Once again, you may want to adjust the current limiting resistor values a bit. Excess current will burn the LED out. If you do not plan on using the LED light on ac, then the diode bridge will make it possible to apply dc power to the LED string without worrying about keeping the input polarity correct. The diode bridge will automatically set the polarity for the LEDs. Both circuits can be assembled on a piece of scrap perf-board. Using point-to-point wiring, you should be able to assemble one in less than half an hour.

If you make up several of these LED lights, try a little experiment. With a very sharp pair of wire cutters, try cutting off the very tip of the LED. This allows the photons to emerge as diffused light instead of the point source light. The plastic tip acts like a lens, so cutting if off helps to spread out the light. You'll have to practice, and you'll end up smashing a few LEDs, but it's worth it.

Checking Out Surplus LEDs

Of course you can purchase ultra bright white LEDs from either Digikey (**www.digikey.com**) or from Mouser Electronics (**www.mouser.com**). You will pay a premium for the parts, but they are well-known companies and ordering is convenient.

You might want to try a Kingbright super bright white 10 mm LED. The Mouser stock number for this LED is 604-L814PWC. It has a typical forward voltage drop of 3.6 V.

Digikey stocks a Lumex white LED with a 3.6 V drop at 20 ma. The Digikey stock number is 67-1604-ND. At $3.75 each, they're kind of pricey.

Lucky for us, we can get ultra bright white LEDs from a number of surplus sources. One of my favorites is All Electronics (**www.allectronics.com**). Their LED-75 ultra bright white LED sells for two dollars each.

Every now and then, All Electronics has a strip of pre-assembled LEDs that go into the third brake light used in today's automobiles. They are ready to go on 12 V dc and produce a blinding bright red glow. See **Figure 2-11**. While I can't give you a specific stock number — they rotate stock all the time — check

Figure 2-11 — This is a surplus brake light. It's full of high brightness LEDs, so it really lights the area up. It's more of a spot than flood so I use it to light up an area so I don't stumble and fall when walking around the shack. This guy sold for about $5 from All Electronics.

their Web site for these and other surplus LED strips. If you can't find the third brake light strips surplus, then a trip to the local auto junk yard would be worth some time.

I came across some rather unusual LED light strips. They are made by Teledeyne. They look like a backlight for some

Figure 2-12 — This is a surplus lighting panel. Heaven only knows what the original use was, but it really lights up the radios. I mount several under the shelf above the radio equipment. They light up white, so you have color retention as well.

sort of display. Perhaps a handheld TV or the backlight behind an LCD display on a gas pump. Who knows? Better yet, who cares? I know that they cost me $5 each on eBay. They operate directly from a 12 V power source. You need to add a 47 Ω, 2 W resistor in series with the power lead to limit the current. I have two mounted directly above my Ten-Tec Jupiter. They provide a nice warm glow that lights up the area completely. Best of all, they only draw 45 mA from the 12 V supply. See **Figure 2-12**.

Another source for surplus LEDs is The Electronic Goldmine (**www.goldmine-elec.com**). Like most surplus electronic dealers, the stock changes all the time. For the best deals, just point your web browser to the sites listed. The stock comes and goes and is in constant state of change. If you find an LED or LED strip that you really like, then you had better quickly order more. They may sell out in a jiffy, never to be had again.

Always keep an eye out for surplus LEDs on the electronics market. I do a Google search every so often on the internet. You would be amazed by what turns up.

COMMERCIAL LIGHTING

Every chance I get, I troll the camping section of a store. During one of my trips I discovered a rather slick LED lantern. Running on a set of four AA batteries, this Eveready lantern will run for 120 hours. You have your choice of longer run time with one LED or twice the light with two LEDs.

Take a trip to your local AutoZone or Pep Boys store. They stock LED replacement lights for use in turn signals and brake lights. Pure white LEDs are harder to find. They're used mostly for side marker lights on autos. If your local auto parts store does not stock them, then point your web browser to **www.4wd.com**. They stock a side marker lamp replacement that is super white. It's their stock number 194LSW.

I've seen ultra bright red and amber lights that will fit inside the RV type light holders. If you have a van, replace the incandescent interior lamps with these LEDs. They really light up the inside and draw a tenth of the power of the old incandescent types.

Another light source I use in my shack on a daily basis

Figure 2-15 — This small light contains white LEDs, the tubes scatter the light making a nice soft glow. Run time is in days.

Figure 2-13 — A Copilot light mounted on the top of a shelf in my shack provides excellent lighting.

Figure 2-14 — This clip on LED light works great for illuminating radio front panels. It runs on a few watch batteries.

I always make it a point to walk through the camping section of the local Wal-Mart. I've seen all kinds of neat stuff hanging on the pegs. The Wal-Mart in my neck of the woods stocks those emergency candles and even chemical light sticks. Be sure to check them out!

Another interesting emergency light is the small clip-on LED lamp shown in **Figure 2-14**. I clip this small light to an edge of a radio cabinet to light up the front panels of an equipment stack. **Figure 2-15** shows a small lantern that has LEDs inside frosted plastic tubes, which help to diffuse the light. This creates a nice soft glow rather than a bright spot.

LIGHTING YOU MAY WANT TO AVOID

I really don't like the idea of using candles for lighting up the shack. They're messy, a fire hazard and just plain old fashioned. If you must use candles in an emergency, make sure you use candles made for lighting. Yes, they are out there. Called "emergency candles," they are made to burn very slowly. They normally have burn times of up to 72 hours per candle. Any large camping store or outdoor supplier will know exactly what you want. See **Figure 2-16**. While you're at it, ask about the special candle holders, too.

If you can't locate "emergency candles" in your neighborhood, try Woodland Products (**www. woodlandproducts. com**). They have them in stock. Another place to shop for long burning emergency candles is

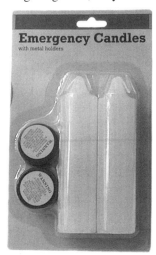

Figure 2-16 — Emergency candles are great to have on hand. However, they do require caution when they are burning.

is the Copilot high output lights. They use incandescent bulbs. Designed for use in police and fire vehicles, they are rugged and good looking. They operate on a 12 V system, making them ideal for use with a battery-based backup system. They are expensive, though. The one desk-mounted light I have in the shack runs about $35. Replacement bulbs are a tad pricey too, about $8 each. **Figure 2-13** shows how I have mine mounted to light up my radios.

Generally, you won't find these at the local Wal-Mart, so search the web for the best prices. A good place to start is Southwest Public Safety (**www.southwestpublicsafety. com**). They offer a large selection at fair prices.

Safety Central (**www.safetycentral.com**).

I also don't care for any of the lanterns that use fuel. They can be very dangerous. They also produce a lot of heat. The fuel must be stored in an out of the way location, never in your home or garage. And of course, if you run out of fuel, you're in the dark again.

EMERGENCY-POWER LIGHTS

Just about everyone has seen these lights mounted on the walls in hotels and businesses. They come on automatically when the grid fails and usually run for several hours on their built-in battery. They come in several flavors. One is a spotlight and it is designed to light up a hallway or exit way. The next is a floodlight. As its name implies, it floods a certain area with light. You see these in supermarkets and other large stores. The third kind are the exit signs hanging from the ceilings in public buildings. The newer ones use LEDs and they will run for almost a day.

If you wan to get an automatic lighting system for home use, you have a choice or two to make. You can get one ready to go from any home center store for about $40 including batteries. Hang it up, plug it in and you're done. **Figure 2-17** shows one model with a special feature: a door labeled "Emergency." Behind that door are several emergency candles. Now that is being prepared!

Another option is to get your hands on a rechargeable flashlight. There are many styles that function as a night light, flash light and emergency light. The run time is usually limited to a few hours. They normally have spot type lenses so they can be used to check the service panel for a popped circuit breaker. **Figure 2-18** shows a Craftsman light I picked up at Sears. It has served me well for a number of years.

If your house or shack uses a modified sine wave inverter, be sure to test the operation of the rechargeable flashlight/emergency light. There are some on the market that will simply melt or be destroyed by the output from a modified sine wave inverter.

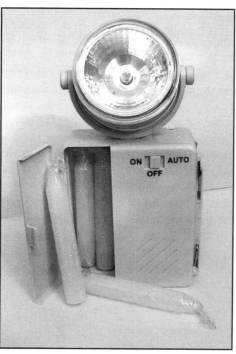

Figure 2-17 — Here's a cheap emergency light. It automatically comes on when the grid power fails. When this emergency light's battery goes dead you can fall back to its built-in emergency feature: Candles!

Figure 2-18 — A good rechargeable spot light is great to have during an emergency. This one came from Sears and sells for about $20.

A homebrew version would be easy to build. You would just need a relay and one of the converted lamps that I will describe in the next section.

An Inexpensive Emergency Lamp

I've kept the best for last. I've used this for years and have always had good luck making them. The best part is they are cheap, good looking and work.

To start this project, you need a broken high intensity light. Any kind any model. Hunt the yard and garage sales for one or two of these. The best ones use a bayonet-style high-intensity light bulb. These bulbs look like a brake light bulb. You may also come across the bi-pin socket used by the halogen bulbs. Either one will work. Avoid any lights that use a standard 120 V ac Edison based lamp.

If your light uses the bayonet style socket, then you can use an exact replacement bulb. The most common problem is that the transfomer's main fuse has opened. This is not a fuse you can readily see once you open the light case. The fuse is nothing more than a short piece of wire that is wound under some tape on the primary side of the transformer. What usually happens is that the bulb burns out. In that instant, the bulb may short, taking the primary-winding fuse with it. Most of the time, it's too much hassle to dig around inside the transformer to fix the fuse wire.

If your light uses a bi-pin socket, then it was designed to use halogen bulbs. They come in three common sizes, 12 W, 24 W and 50 W. Be careful not to exceed the lamp rating. The heat generated by the higher wattage bulb may cause plastic parts to melt or the entire lamp housing to get way too hot.

If you are using halogen bulbs, keep an eye on the current the bulbs demand. Assuming a 12.5 V battery backup system, a 12 W bulb will draw just under one amp. A 24 W bulb will draw twice that much current. The 50 W bulb will demand 4 A from the battery! It should be clear to everyone that a 4 A load can impact the run time of your battery bank.

It would be best to seek

out an old lamp that uses the bayonet style bulb holder. This would be an ideal location to install those LED replacement bulbs. Converting the light over to 12 V dc is child's play. Here is how you do it.

First, cut off the 120 V ac plug. Leave the original power cord as long as you can. Next disconnect the bulb's socket from the transformer. *Do not* remove and throw away the transformer, even if the one you have is cooked. The transformer is the ballast that holds the light on the table. Now, connect the bulb's socket wires to the power cord using an inline ATC fuse holder. Insert a suitably rated ATC fuse into the holder. If you using a 12 W bulb, then a 1.5 A fuse will be fine. Likewise, smaller LED bulbs will require smaller fuses.

Although I like to hide the fuse inside the lamp base, replacing it then requires opening the lamp case. An inline fuse holder could be added to the power line outside the base. If you do this, install the fuse holder as close to the lamp base as possible. You want to protect the supply cord from melting in the event of a short circuit.

On the business end of the lamp, I install a set of Anderson Power Pole® connectors. That way, I can move the lamp around and simply plug it into one of my low voltage outlet panels. If you don't use the Anderson Power Pole® connectors, by all means you MUST cut off the 120 V ac plug! That's it! Now you have a 12 V dc light that will run either from your current station power supply or from the emergency backup batteries.

Outdoor Lighting Comes Inside

If you're looking for another way to light up your shack, drop by the outdoor lighting section of your local home center. You want the type of light holder that looks like a spotlight. Standard outdoor lighting fixtures just throw out light in all directions. They produce way too much glare, making them unsuitable for use in the shack.

Get the metal kind rather than the plastic body. The metal one can be installed directly onto metal conduit. Installed on the ceiling, they can provide task lighting. Since these lights run from a 12 V system, they are easy to convert over to our backup battery system. Again, keep an eye on the total current drawn by these lights.

Recently, outdoor lighting has moved towards ultrabright LEDs instead of incandescent bulbs. I have a small solar powered LED outdoor spot light that will light up the entire side of my house.

Why not just use the 120 V ac lights? They require good ol' 120 V ac, so you would need a dc to ac inverter to run them. And the current draw can sometimes be rather large. As a matter of fact, although I have more than enough inverter capacity and battery capacity, I have chosen to use low voltage lighting in the event of a power emergency. The fluorescent lighting in my shack would simply draw too much power from the battery bank. It's better to read by the light of an LED lamp and communicate than to have blinding light for a few hours and not be able to talk to anyone.

As you can see — and no pun intended — having emergency lighting when the power goes out is very important. Even if it is just a little bit of light in the darkness, everyone will feel safe and secure. When hurricane Charlie tore through south Florida during the 2004 hurricane season, a few LED lights would have been a godsend.

Chapter 3

Solar Power

With the possible exception of geothermal power, all energy on this planet originally came from our Sun. Sunlight is made up of tiny energy packets called *photons*. Every minute, enough of this energy reaches the Earth to meet the energy demand for the whole world.

The oil, coal and natural gas in the ground — and even the wind — are all directly connected to the Sun. Crude oil is really nothing more than dead plant and animal matter compressed under tremendous pressure and heat. Cook this mixture for a few million years and you end up with coal, crude oil and natural gas. But the plants first took the Sun's energy and converted it into complex hydrocarbons. Then the animals consumed the plants, died off and got mixed in with the other plant material. In some ways, the gasoline you put in your car is really nothing more than dead dinosaurs!

You don't have to wait a million years to use the Sun's power. In fact, it's quite easy to do. Believe it or not, it's not all that expensive anymore. Sure, you can't disconnect your house from the grid when the local power company is selling you juice for five cents per kW, but solar power can be cost effective. It's like water in the desert. What price do you want to pay to have power when everyone else is sitting in the dark? It simply amazes me that a ham will spend $15,000 for an HF rig and not $1000 for a solar-electric power backup system.

SOLAR ENERGY FOR EMERGENCY POWER

Solar panels can be used as the basis for supplying energy at any location on the planet. They require no fuel other than sunlight. They produce no emissions, no fumes and have no rotating parts. They are completely silent. Other than an occasional wipe down to remove dust or dirt, they are virtually maintenance free.

One of the best aspects of solar power is the ability to expand generating capacity as your needs grow and your budget allows. You can start off with a 40-W panel and add on to it at a later date. And believe me, you *will* add on. Running your shack, even just for the fun of it, from solar power is very addictive. Once it gets into your blood, you're hooked.

We use solar electric systems with batteries to store the power the solar panels produce during the day for use at

These dual axis trackers hold eighteen 75 W Siemens modules. There are three trackers. The array is wired for a nominal 24 V system.

night. Because the batteries act as reservoirs, collecting the charge for later use, we don't need to use large solar panels. A small solar panel can keep the battery bank fully charged with relatively little effort.

For instance, have you ever seen a small solar panel on the roof of a lakeside cabin and wondered how it could possibly power all the appliances inside? The answer is simple: the solar panel charges the battery all day long, all week long. The cabin is only used on the weekend, so the battery is always charged and ready to go. After the battery has been discharged, it has another week to recharge.

This small "solarverter" can operate quite a few personal electronic devices.

TWO TYPES OF SOLAR ELECTRIC SYSTEMS

While the majority of systems use batteries to store the energy produced during the day, there are some instances where you can use solar electric directly. A notable example of this is a solar-powered fan. When the Sun shines, the fan runs. It's a simple as that.

There are some smaller solar panels being sold that will run radios, CD players and tape recorders directly from the Sun without batteries. The solarverter shown in the accompanying photo is a perfect example of this. The small solar panel has the correct voltage and current to operate a large number of small dc loads directly from the Sun. Of course, the only drawback is that when the Sun is behind the clouds, the load quits. Some small radios will operate with this solar panel on brightly overcast days, but for maximum bang, you need plenty of direct sunshine.

If you wanted to operate your radio equipment directly from a nominal 12 V solar panel, you need to add a series voltage regulator. As you will read later on, the peak power point of the solar panel is way too high for safe operation when connected directly to our loads. If you were to connect a solar panel directly to your ICOM 706 transceiver, for example, 10 seconds after the Sun pops out from under a cloud you'll be calling 1-800-FIXME. The peak voltage from the solar panel would be near 17 V — way too high for the little ICOM radio to handle.

SOLAR POWER BASICS

Photovoltaics. The name says it all. The word *photo* meaning light, *voltaic* meaning electric. So, the word photovoltaic (PV) literally means electricity from light. A solar cell converts light into electricity.

What makes a solar cell work? It's really quite complex. We're talking atoms and single electrons all moving about with a few photons thrown into the soup. This complex dance takes place within the first few microns of a silicon wafer that has been specially doped with certain chemicals. When all is said and done, exposing the cell to light produces an electrical current that flows between the top and bottom layers of the silicon wafer. The workings of a solar cell can be summed up as moving electrons.

A closer look into this electron movement involves the light particles called *photons*. When sunlight hits a photovoltaic cell, photons move into the cell. When a photon hits an electron it dislodges an electron leaving a "hole" in its place. The loose electrons then move toward the top layer of the cell. The process continues as long as light strikes the cell. Electrons are dislodged and holes are created.

The top half of the photovoltaic cell is negative while the bottom half is positive. The material that separates the two layers is called a P-N junction.

The loose electrons move out the top of the cell via a thin grid of metal called the *collection grid* and into the external electrical circuit. Electrons from further back in the circuit move up to fill the empty electron holes. As long as light continues to fall upon the cell, electrons will move through the external circuit.

Look at this way. If you park your car in the sunlight, the roof will begin to heat. This heat is due to the action of the Sun exciting the electrons within the molecules of steel. The heat generated is cause by friction produced by the electrons moving about. Add a way to collect these moving electrons and you have a solar cell. This is a very simple explanation of how a solar cell works.

SEVERAL TYPES OF SOLAR CELLS

Today there are four basic cell types being produced:
- *Single cell*
- *Polycrystalline*
- *Amorphous*
- *Ribbon*

There are a few other types produced too, but they are mainly used in spacecraft and therefore extremely expensive. For us on this planet, the four examples above will be most common.

The Single Silicon Cell

This is by far the most popular type of solar cell produced today. It is also the cell that has been produced the longest. It has a great track record.

Refining silicon in a furnace produces the single cell. The

You don't need to start out big. Solar power is the one source of power that allows you add on as your needs grow and your budget allows. This array consists of four Solarex MSX-77 modules.

silicon is melted, purified and allowed to crystallize into solid silicon ingots. The exact recipe used to melt and purify the silicon is a guarded trade secret. After the silicon has cooled, the ingot is sliced paper thin using either lasers or wire saws. The silicon is cut into wafers about 200 to 300 microns thick. The wafers are treated to several chemical dopings and finally a grid pattern is applied to the top to collect the electrons. This is a condensed version of how to make a solar cell. There are really a dozen or more steps that the wafer must go though before it becomes a working solar cell.

The single cells have a uniform blue or black color. They appear shinny because of an anti-reflective coating applied to the cells.

There's also a very fine web of connecting grid lines to collect those loose electrons. There is a fine line between making this collection grid large enough to get all the electrons without blocking the photons that produce the loose electrons in the first place. It's fine balancing act each company must deal with to improve cell efficiency.

The Polycrystalline Cell

The polycrystalline cell is produced the same way, but the recipe is a bit different. A cheaper grade of silicon is generally used, resulting in a slight price advantage over the single crystal ingot. The cheaper grade of silicon also makes for a cell that is slightly less efficient. In the end, the two seem to balance each other. The result makes the polycrystalline cell appear frosted. Some people describe them as frosted ice on a window pane.

Single cell and polycrystalline solar panels come with 25 year warranties. Some manufacturers offer longer warranties.

The Amorphous Cell

The amorphous cell is made using a process quite different from what we have discussed so far. Most amorphous cells are manufactured by coating a suitable substrate with a plasma. An amorphous cell is usually only 20 microns thick. Current technology limits us to sawing silicon ingots no thinner than 200 microns. Since the conversion of sunlight to electricity is done within the first several microns of silicon, there is a tremendous savings on material when amorphous cells are produced. Because the active material is so thin, amorphous cells have taken on a second name: *thin film*. So when you run across the term "thin film," think amorphous cells.

There are a few more tricks the thin film cell technology can do. Since the thin film is so...well...*thin*, the material can be deposited on glass, making the glass a translucent solar panel. This not only provides shade, but generates

electricity at the same time. You see a lot of this technology used in buildings.

The other trick is that each layer of silicon can be "fine tuned" to a specific wavelength of light. With ultra thin layers, multiple layers are deposited one on top of another.

When introduced in the mid to late '70s the Arco 16-2000 was a standout product. Its name reflected its power: 16 V at 2000 mA. Notice these panels only have 33 cells. While not a "self regulating" panel, it was a cost cutting design. The higher-power panel was the 16-2300. They set you back lots of bucks in 1978!

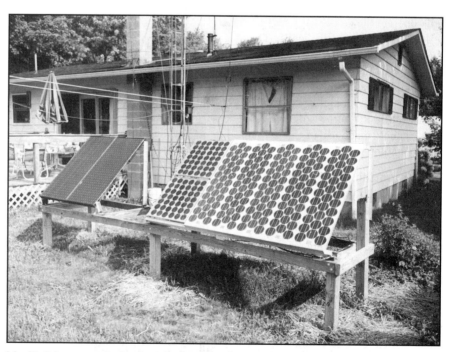

I built this as part of a test bed to monitor various modules. On the left you see two Sovonic R-100 thin film modules. In the middle, there are two original Solarex modules. The last four are Arco 16-2000 modules.

It doesn't matter if the Sun is charging up batteries in these solar lights or charging the battery in your QRP rig with this old Arco module — they all work the same way.

Each layer is sensitive to a particular wavelength, producing a solar cell that is more efficient. Current thin film manufacturers are using up to three layers of active material.

The Ribbon Cell

The last cell type is called a ribbon cell. They are made by growing a ribbon of molten silicon and pulling it though a die. These cells operate the same way as single cell and polycrystalline cells. The anti-reflective coating used on these cells gives them a prismatic, rainbow-like appearance.

You might come across some older Mobile RA series solar panels. These panels made use of the ribbon cell technology.

Comparing the Four Cell Types

Both the single cell and polycrystalline cells (including the above ribbon cells) convert between 12 and 16% of light striking their surface into electricity. The single cell has a slight edge in conversion efficiency. Both the single cell and polycrystalline cells cost about the same to produce, with the edge going to polycrystalline cells.

Although the amorphous cell, with its dramatic reduction in mate-

rial, should have a price advantage, so far that has not appeared in the real world. However, the amorphous cell does come in with a slight advantage in price over the other two technologies.

The amorphous cell does suffer from lower efficiency. Current amorphous technology runs about 6% efficient in converting light into electricity. Put a different way, a 50-W solar panel made of amorphous cell technology will be twice as large as a similar single cell or polycrystalline cell panel.

Since you can deposit the amorphous cell material on just about any substrate, amorphous cells can be made on plastic, glass or even stainless steel. The result is a solar cell that is resistant to damage. That's a big advantage if your cells are subjected to harsh climates and kids with rocks. The Unisolar flexible solar panels are a result of this technology.

Another plus for the amorphous cell is its ability to be manufactured without glass. Amorphous PV cells can be very light weight.

The amorphous cells have one more oddity up their sleeve: they suffer from power loss when exposed to sunlight. A Unisolar US64 panel is rated at 64 W out of the box. When first exposed to sunlight, the panel may produce up to 75 W. After the first 90 days or so, the power will drop down to 64 W and stay there for the life of the panel. The power will *not* decrease further than the rated power of the panel, though. Current third-generation amorphous solar panels have 20 year warranties against loss of power.

So, which cell type is best for a ham radio back-up system? Well, there's really no junk out there anymore. The guys that were making junk have sold their wares and moved

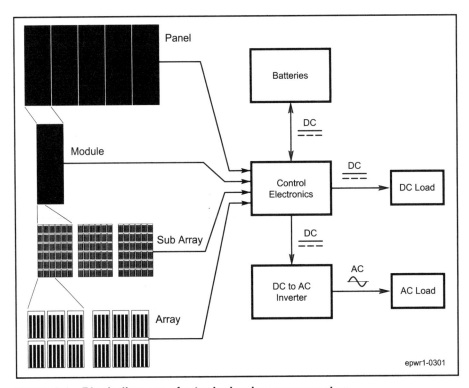

Figure 3.1—Block diagram of a typical solar power system.

on to the next fad.

Solar cells by BP Solar, Astro Power and Shell Solar are all first-run single silicon or polycrystalline cells. For amorphous, the front leader has always been Unisolar. I'll present a rundown of various companies and their PV modules later on in this chapter.

You may have noticed I have mentioned only individual cells. Once cells are made, they are assembled into modules. Two or more modules form a *panel*. Two or more panels form an array.

Through the years, the distinction between module and panel as become blurred. I interchange the name module and panel all the time. Most people think of a string of solar cells connected together in a frame as a "solar panel."

EFFICIENCY AND DOLLARS PER WATT

When I have given talks on solar power to groups of ham radio operators, without fail, someone will ask when cell efficiency will improve.

Right now, cell efficiency is only important if you plan on putting a satellite in space. In that application, you must have the maximum efficiency our technology can provide. For those of us with our feet planted firmly on the ground, efficiency is not the key word. Dollars per watt *is*. Okay, that's three words. The bottom line is this: I really don't care if the solar panel only converts 2% of light into electricity if the cost per watt is 28 cents! Right now, if you purchase a large quantity of panels, you can push the dollars per watt to somewhere near $4. Some solar cells used in spacecraft push the efficiency to over 35%. For me, I would be thrilled at 2% with a dollars per watt cost of 28 cents!

PHOTOVOLTAIC BASICS IN REAL LIFE

Now that you have learned how solar cells are made, it's time to start putting them to work. One solar cell is not going to run much of anything. However, we can add cells in series for voltage and add cells in parallel for current.

Both of these solar cells produce the same amount of voltage — about 0.5 V each. The only difference is the amount of current produced. The larger cell can provide over 5 A, while the smaller one is lucky to supply 50 mA.

From this basic building block, we can add panels to increase voltage and add panels to increase current. Whole arrays may be combined to increase operating voltage and current even further. Some solar panels have voltage ratings up to and including 600 V dc. See the block diagram of a typical solar electric system in **Figure 3.1**.

Increasing operating voltage

No matter if the cell is made of single silicon cells or polycrystalline cells, when exposed to light, they produce about half a volt. The amorphous cells produce slightly more

Who is Who?

Get your program here! Get your program here! You can't tell the players without a program!

Yes, it really has become so complicated that you need a program to follow what is going on in the solar electric industry. The names of the players keep changing with the times.

The oldest names are long gone. Arco Solar was the granddaddy of them all. Arco and Solarex were the two big players. The Arco solar company was located on the West Coast while Solarex was based in Maryland.

Arco was first purchased by Siemens. Together they became known as Siemens Solar. All the while Solarex was carving out a niche for itself with its polycrystalline cells, Siemens Solar sold single crystal cells.

Right before the year 2000, BP Solar started to get a toehold in North America. They went after market share and took a nice slice of pie from the two major players. Several years ago, BP Solar purchased Solarex. The company was called BP Solarex, but is now just known again as BP Solar.

Meanwhile, back at the ranch, Siemens solar was purchased by Shell Energy. It's now call Shell Solar. Try to say that one ten times fast!

Astro Solar Power started out as a company that just made solar cells. The cells were sold to anyone that wanted to assemble them into panels. A lot of OEM and custom made solar panels have Astro Solar Power cells in them.

Astro Solar Power then decided to start their own line of solar panels. They were quite successful — so much so that they grew too fast. Before someone turned the light out at the end of the tunnel, Astro Solar Power went bankrupt. They are now owned by General Electric.

Sovonics Solar later became Unisolar. They are still based out of Troy, Michigan.

There are several new players, too. Sanyo, Sharp and Kyocera are now in the North America market. While new to us here in North America, these companies have excellent track records throughout the world.

Several companies in India are now shipping panels and cells into North America as well. These show up as special OEM version panels.

The imported solar panels generally have the same 20+ year warranties as the ones made in North America. At the time this was written, there was no "junk" being sold. The guys that marketed the junk panels have now moved on to beanie babies and low carb foods.

voltage. It makes no difference what size or shape the cell is. The output voltage will be about half a volt.

To produce a useable voltage we need to add cells in series. Most solar panels have 36 series cells. Together they produce about 18 V. Yes, that's too high for us to use, but I'll explain the reason for that value later.

Increasing Current

The amount of current produced by a single solar cell depends on two factors. One is the physical size of the cell. The larger the cell, the more current it will produce. Current technology limits the single crystal and polycrystalline cells to around eight inches in width. Also note that the larger cells are not perfectly round. Their corners have been cut to allow increased density when mounted in a frame.

The other factor is the amount of light hitting the surface of the cell. In solar speak, that is called *insolation*. The standard insolation is measured as 1000 W/m^2 irradiance at 25°C temperature. I'll mention this several times. It's an important number to know when working with solar panels.

A HOMEBREW SERIES REGULATOR FOR SOLAR DIRECT OPERATION

If you think about it for a second or two, all you need to operate "solar direct" is a regulator between the solar panels and the load.

But let's discuss that load. If you're thinking of running a 100-W transceiver, think carefully. A run-of-the-mill 100 W transceiver requires about 18 to 20 A to operate at full output. Considering the current rating of an average panel, we're talking about a lot of solar panels! So, if you want to run solar direct, without any battery storage, think lower-current loads. The low-power ICOM 703 transceiver would be a good choice, as would the SGC-2020 or the Ten Tec Argonaut V. Any 2-meter transceiver in the 10 to 15 W range would be okay, too.

As for the solar panel, if your rig requires 4 A at full transmit output, you will need a solar panel that'll produce that much power under full sun. Let me be clear here: the ratings on solar panel labels are for power generated under almost *perfect* conditions.

The Circuit

This is a bare bones, easy to build circuit. The schematic for a regulator is shown in **Figure 3.2**.

You can use either one of the 78XX series regulators for up to 1 A of current, or use a higher-rated device like the

LM350STEEL, which will do 3 A. If you wanted to, an LM317 adjustable regulator would allow you to select a different voltage to allow you to directly operate loads that require different voltages, as shown in **Figure 3.3**.

A circuit that uses a transistor and zener diode is shown in **Figure 3.4**. This circuit lacks the precise voltage regulation that the 78XX series of IC regulators provide, but it uses junk box parts all the way. Also, the 78XX series regulators have onboard short circuit protection. If you build the circuit using the transistor, a short circuit will destroy the device faster than a fuse could protect it. So, to keep things from going up in smoke, I usually add a very low value resistor to limit the current. The 0.33 Ω, 2 W resistor will take the current for a few seconds and keep the transistor happy until the short is removed. It's a brute force method of short circuit protection, but it does work in a pinch.

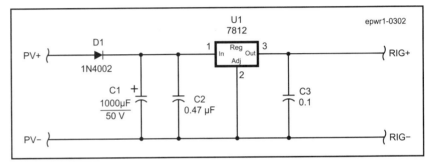

Figure 3.2—A simple solar direct regulator using a 78XX series regulator.

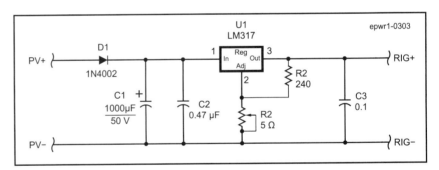

Figure 3.3—The same basic circuit as Figure 3.2, but this time with an LM317 adjustable regulator for devices that need another voltage besides 12 V. (Usually personal radios, tape players and so on.)

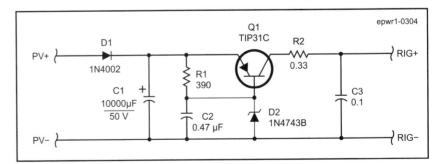

Figure 3.4—This simple circuit is totally junk box. The PNP transistor is regulated by the zener diode. The 0.33 Ω resistor provides some short circuit protection for the transistor.

In all cases, there is a diode in series on the input side of the circuit. This prevents the solar panel from trying to charge the input capacitor if the panel is connected backwards.

Notice the high input filter capacitance. This acts as a storage tank to keep your load running while the Sun plays dodgeball with the clouds. Deciding on an exact value is between you, your load and the solar panel you have. Too much capacitance and too little solar panel means you spend a lot of your time just charging the capacitor. Values between 10,000 μF and 100,000 μF are normally seen. Because the solar panel can produce upwards of 22 V open circuit, the input capacitor must be rated at least at 50 V.

Build the circuit on either a small printed circuit board or on perf-board. Whatever regulator you use, it will get hot because it will be dropping a lot of voltage. A heat sink is a must. You should then assemble the circuit in such a way as to prevent it from damage when you take it outside to operate your gear.

Solar Panel Power Ratings

Remember the 18 V we had in our string of cells? We all know that is too high for the 12 V rigs we run. Why then would one want a solar panel that produces 18 V? The answer is called the *power point* of the panel.

With nothing connected to your solar panel but your digital VOM, the meter will read about 20 to 22 V. This is known as the *open circuit voltage*. That's the correct open circuit voltage for a panel rated for a nominal 12 V system.

When we start to draw current from the solar panel, the voltage drops in an initial downward spike. As we increase the current drain, the voltage continues to fall. Soon we reach a point, known as the *voltage knee*, where the panel is generating its maximum power. Increase the cur-

rent drain a tad more and the voltage will drop to the power point where the panel is producing maximum voltage at maximum current. An example of a power plot is shown in **Figure 3.5**.

Most panels produced today have an operating voltage of 17 to 17.5 V. The power in watts for a panel is the maximum current produced at the operating voltage of the panel, times that operating voltage. For example, let's say that our solar panel is supposed to produce 3 A at 17 V under test conditions. The panel would then be rated at 51 W. It's important to note that the panel is not rated at 12 V at 3 A even though the panel is a "12 V" panel.

This brings up an interesting point. In an application where the solar panel is charging a battery, the operating voltage of the panel is effectively determined by the battery voltage. This varies over a narrow range depending on state-of-charge of the battery and battery ambient temperature. The operating voltage is usually between 1 and 4 V *lower* than the voltage at which peak solar panel power figures are generated. Luckily for us, in real world conditions, the current capability changes very little from the peak power voltage of 17 V to our nominal system operating voltage of 12 V.

It should be now be clear to you that as the battery bank becomes fully charged, the peak-power voltage will become higher, approaching the 17 V peak-power voltage of the panel. This is why you need a charge controller between the solar panel and the battery under charge. I'll talk about charge controllers in more detail in the next chapter.

For crystalline silicon solar panels, the operating voltage will decrease about one half to one percent for each degree Celsius temperature rise. That's why on a module's specifications you'll see temperature plots and voltage at 25°C, 20°C and 50°C. Bottom line? The hotter the solar panel the lower its operating voltage. If you get the solar panel hot enough, the voltage will drop to the point that current is affected. The lower voltage can't charge the battery.

A number of years ago, a few companies came up with a clever idea: They decided to remove several cells from a module. That lowered the peak voltage by several volts. When the battery was discharged, the panel produced enough voltage to charge the battery to capacity. When the battery voltage increased, the panel's peak voltage was low enough to prevent overcharging the battery. The result was a "self regulating" panel. They sold a lot of them — until a problem began to show up. If you put these panels on a roof of your cottage by the cool lake, everything worked as planned. But if you installed them on the roof of an RV and went to Arizona, the heat from the midday sun would overheat the panels and bring charging to a grinding halt.

These self regulating panels are easy to spot on the used market. Just count the number of cells you see in the module. Modules with 30 to 32 cells are considered self regulating. If you happen to find some self regulating modules, save your money (unless it is a *real* bargain price).

At the other end of the scale, you can special order modules with 44 cells. These high voltage modules are al-

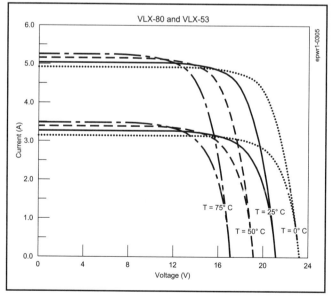

Figure 3.5—Typical current/voltage plots for a solar module. These show the plots of the Solarex VLX80 and VLX53 modules at several temperatures. Notice how much more voltage the panels produce when the temperature is lower.

most never used to charge batteries. It's like going down the road with the pedal to the metal with the emergency brake on. These panels are used with peak power trackers connected to dc water pumps. They also show up in larger grid intertie systems where the system voltage may approach 600 V dc. Although this type of panel is also rare to find, I would choose it over the self regulating module.

Because heat does drop the voltage of the panel, it is important that the panels never be placed directly on a roof. You should mount the panels so that there is at least a 2-inch air space between the roof material and the solar panel.

STANDARD TEST CONDITIONS (STC)

It's always a good idea to compare apples to apples. When the solar industry was young, they all got together and came up with what is now called the standard test conditions. Basically, every manufacturer would test and rate their panels by the same test everyone else uses.

That means when you purchase your solar panel and it's rated at 75 W, you get 75 W under *standard test conditions*. So, just what are the standard test conditions? There are two factors that determine how a solar panel is tested:
• The amount of light striking the panel
• The temperature of the panel

Like I mentioned earlier, solar cells don't like hot weather. The hotter the cell, the less voltage it will produce. The standard test condition for temperature is 25°C.

Do you remember that I said the amount of current produced is determined by cell size and amount of light striking the surface? This is where the 1000 W/m² irradiance at 25°C temperature is used as the standard insolation. When you take the 1000 W/m² irradiance at 25°C temperature you have what is called "one sun." One sun conditions don't happen that often. In fact, you may never notice it unless you have the instruments that can measure the solar irradiance.

Normal Operating Cell Temperature

This is a figure you see every now and then on a solar panel's label. The normal operating cell temperature (NOCT) gives the estimated temperature of a module when operating under 800 W/m² irradiance, 20°C ambient temperature and wind speed of 1 meter per second. NOCT is used to estimate the nominal operating temperature of a module in its working environment.

MEASURING SOLAR POWER

While most of us drag out a multimeter to measure what is really going on with Mr. Sun, you really need a widget called a *pyranometer*. The pyranometer measures both the direct and diffused components of sunlight. By integrating both of these values, an estimate of insolation may be generated. A lab grade pyranometer can set you back thousands of dollars, with the "cheap ones" run anywhere from $40 to about $100. These pyranometers use a calibrated solar cell. The current produced by this calibrated cell is measured on a small analog meter. Read the scale directly read in 1000 W/m² irradiance. Half scale would mean you have 500 W/m² irradiance. The cheaper pryanometers do not take temperature into consideration.

The Solar Cell Turns 50

The year that I was born, 1954, Bell Labs press releases made a bold statement: "Since it has no moving parts and nothing is consumed or destroyed, the Bell Solar Battery should theoretically last indefinitely." Well, 50 years is not exactly forever, but so far, so good. I am still around and so are some of the original cells.

One of the original cells, which was about 6% efficient in 1953, now has a conversion efficiency of about 1.4%. The cell is not encapsulated and has taken a beating during the last five decades, as evidenced by one end having been chipped off. A 1955 Bell cell, however is still going and going and going. This one is encapsulated and is still a thing of beauty and strength. And, 49 years later, the original 6% cell boasts an NREL-verified efficiency of 5.1%.

Although not really out of reach, most of us will never own a pryanometer. You can get a fairly good idea of how much sun you're getting at any given time by just looking at the ground. That's right, look at the ground for shadows, not at the Sun! Look for a shadow that comes from a telephone pole, sign post or even a tree. Now, don't look at the center of the shadow, but look at the edges. The more sunlight, (and less goop in the air) the sharper the edges will be. Particulates and other stuff in the air will soften the edges, because there is less light making the shadow. I am not a walking pryanometer, but with 30 years of experience, I can make fairly close estimates.

UNDERSTANDING SOLAR MODULE "LABEL SPEAK"

On the back of most solar panels you'll see a label giving the specifications for that particular model of panel. Solar panel manufacturers generally name a solar panel by the wattage it produces. A Unisolar US64 produces 64 W. A Shell SM110 produces 110 W and so on. Besides the wattage of the panel, there will be other solar techno babble on the label. Here's what they mean.

V_{mp}: This is the Voltage at maximum power. Usually between 17 and 18 V. Sometimes you will see this same number shown as V_{pp} or Voltage peak power.

I_{mp}: This is current at maximum power. This number varies with the amount of current produced by the solar panel. Sometimes this same number will be shown as I_{max}. Or simply current at maximum. Some manufacturers call it maximum output power, just output, peak power, rated power or other terms.

I_{sc}: This is the short-circuit current of the panel express in amps.

V_{oc}: The open-circuit voltage represents the voltage of the panel when no load is applied. Most silicon crystalline panels will produce between 20 and 23 V.

STC: That's our 1000 W/m² irradiance at 25°C temperature test reference. You will sometimes see on the label that "I_{mp} at STC." Which means the current power point was generated at stand test conditions.

When a manufacturer produces a solar panel, a 75 W module is designed to produce 75 W under STC. The module you have in your hands may produce more, it may produce less. The average rating on that module, however, will be 75 W.

When Solarex was still Solarex, they flash tested each module. Each solar panel that left their factory had a label fixed to it that showed exactly what that panel would produce.

Here's how they did it. Each solar panel was connected to a computer. The solar panel would then be "flashed." The flash, not unlike that of a camera flash, would reproduce that 1000 W/m^2 irradiance and in an instant the computer would calculate the power point of the panel. So if you held in your hands an MSX64, you knew without question that it produced 64 W at STC.

This precise labeling was a great idea until the marketing people got hold of it. Every time a batch of new silicon was produced it would produce a different outcome. Some modules would flash 53 W, some would flash 50 W and then some would produce 64 W. So, what you ended up with was a bewildering amount of different module models. The 53 W modules became MSX53, 50 W became MSX50, 64 W became MSX64 and so on. There were VLX53 and VLX58 with the same physical sizes, but slightly different in their output. They also had a slightly higher price, too.

Since it was hard to produce the same chemistry on each batch of silicon, it was nearly impossible to reproduce a run of, let's say, MSX53 modules. You might end up with MSX64s or MSX58s.

Because of the flash testing, solar panel dealers like me had the option of ordering "hot panels" that flashed out with a higher V_{pp} for use in the southwest. The higher voltage allowed the panel to work better under the hot Arizona sun.

AMP HOURS AND SOLAR PANELS

Although we normally talk about watts when we purchase our panels, the end result is measured in amp hours. Because batteries speak in amp hours, it's best that we keep both the solar panels and the battery talking on the same page.

When you are ready to purchase a new or used panel, the first thing out of your mouth will be, "How many watts and how much money?" You probably never ask "What is the amp hour rating of the panel?"

We do this all the time without thinking. It works the other way too. You don't ask the guy selling the 2 meter rig what the output current is. No, you ask how many watts the thing sends to the antenna.

Since solar panels are rated in watts at the standard test conditions, STC, speaking in watt hours can be misleading. By using amp hours we get a better real-world feel of exactly how much power a given solar panel will produce. Let's look at an example.

The term watt hour is just that, the amount of watts produced in one hour. So under STC a 75 W panel will produce 75 W in one hour. If your location has an average of 6.5 hours of sun per day, then this panel will produce

What Size Panel do I Need?

There is a complete chapter later on in this book that deals with load sizing, but I still should take time in this chapter to discuss the number one question: "What do I need to run my radio?"

Doing a load sizing calculation is one thing, but most of us size our solar panels by the amount of money in the checkbook. There are some rules of thumb that I've found that work fairly well.

For home stations, don't even think of using anything less than a 32 W panel. Fifty watts would be better yet. The most popular solar panels are now rated at least 75 to 80 W. If you limit your radio use during the weekends, a 32 W panel can produce enough power for casual operating, if you have enough sunlight.

Portable stations such as those used on camping trips or during field day can be powered from a 10-W panel in most instances. A 10-W panel is about the size of an open *QST* magazine. That's just about as big as I would want to carry around on my back. The most popular panel for this is the BP/Solarex MSX-10 Lite module. It's made without glass and is unbreakable under most conditions. You can walk on it and it will stand up to just about anything smashing against it. It is *not* bendable or foldable. The slick thing about this panel is you can strap it on the backpack frame. With your battery pack nestled inside your backpack, you can recharge your battery as you hike up the trail.

If you need more power in a portable application and don't want to mess with a glass panel, then BP/Solarex also makes a 32 W version of this Lite panel, but at this size, portability becomes an issue. If size and weight are a problem with portable operation, then BP/Solarex also has a 5 W version of the same panel.

487.5 watt-hours per day. Remember, this figure is based on STC and in the real world will vary a great deal. To complicate matters, the wattage of our panel is based on peak power. The power point of a photovoltaic system is determined by the battery voltage, so the power point may not be the same as peak wattage. My solar array is rated at 2.5 kW peak. The operating wattage is much lower. The value is determined by the battery bank state-of-charge.

Because we used watt hours to rate the panel, the actual amount of power getting into our batteries may be quite less than we expected. This is because the wattage is based on the peak power produced by the panel. This is why we use amp hours when dealing with powering loads and charging batteries. When using amp hours, we get a better handle on what is really being produced and sent to the battery bank under charge.

An amp hour is exactly that, one amp for a period of one hour. In the example above, our panel under STC is producing 487.5 watt hours peak. When charging the battery bank, we get 28.67 amp hours per day, per panel, using the same 6.5 hours of sun per day. How did I get that figure? It's simple; take the I_{max} rating of the panel times the hours of sun per day. In the 75 W example above, the label states that

Solar panels really like the cold weather. This array may produce more power on this bright, cold winter day than under standard test conditions.

This homebrew mounting system consists of treated lumber and commercial solar panel mounts.

A strange grouping of solar panels. This eco village is set up to demonstrate various solar electric systems and balance of components.

this panel will produce 4.41 A at 17 V. That's 75 W. If this panel sits in the sun at STC for one hour, we get 4.41 amp hours. So, 6.5 hours times 4.41 amp hours is 28.67 amp hours per panel per day.

Even this figure of amp hours can be misleading since it is based on the I_{max} of the panel. Once again, the battery bank will set the tone of how many amp hours the battery requires to recharge. When the battery bank is nearly full, the amp hours being generated will not be the same as what the panel can produce.

Because all the figures generated in the example above use the STC of 1000 W/m^2 irradiance at 25°C, it does not reflect an accurate real world figure. To get a better real world figure, take the 28.67 amp hours and remove 30% for a real world figure of 20 amp hours per day. Amp hours become an important parameter when we begin sizing our solar panel system to our loads.

USING SOLAR PANELS IN THE REAL WORLD

With all this techno babble going on, it's time we look

at how a solar panel works in the real world. After all, you just want to keep your station up and on the air when the power goes out.

Some Rules of Thumb

I know everyone hates rules. I realize the more specific the rule, the greater the number of exceptions.

I've said it before, but the standard test condition of 1000 W/m^2 irradiance is how the *manufacturers* rate their panels. In the real world, we hardly ever see that amount of sunlight. Most important is the 25°C rating. Unless you're on a mountain or in North Dakota in the middle of January, your solar panel is not going to be that cool. The heating effects of the sunlight will raise the temperature of the solar panel to be significantly hotter than the surrounding air temperature. You would burn your hand if you place it on the surface of the solar panel, even on a cold winter day! Bottom line? In the real world, figure on 800 W/m^2 irradiance of insolation, and generally at a much higher module temperature.

Taking into account all the junk and goo in the air, along with heating of the panel in the August sun and you're looking at a real world output of about 70% of rated power under standard test conditions.

The amount of junk in the air really affects how our module will perform. On those hazy days of August the amount of pollution, ozone and just plain old dirt can really take a toll on the panel output. When our county is placed under an ozone alert in the summer, I can watch the current

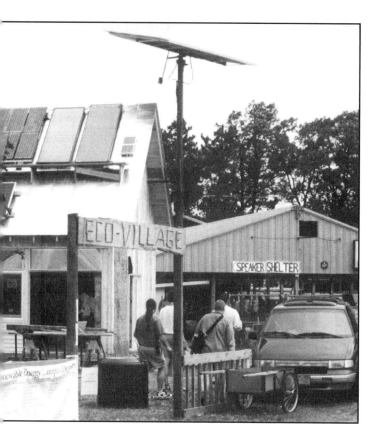

This small 10 W panel is perfect for portable operation. This panel is at least 15 years old and still produces rated power. Newer panels made for portable use have stainless steel back panes and high tech coatings.

drop from my solar array as the midday sun starts cooking the air.

SHADING YOUR PANELS CAN CAUSE DAMAGE

Solar panels don't like any kind of shading. Not even the shadow of the coax from your dipole. That's just enough to cause the current to stop flowing. Never allow any shadows to fall across your solar panels if at all possible. The shading of just a small portion of a solar panel can greatly reduce its output. It's even possible for a shadow to destroy a solar panel in a high voltage system. I'll explain.

Most of us will use a nominal 12 V system. So we normally just parallel modules to produce the required current for our loads. On the other hand, if you need to increase system voltage you add modules in series until you reach your operating voltage. In a nominal 24 V system, two modules would be wired in series. In my 48 V system I have four modules in series.

Solar modules in series must carry the same current. If one or more of the cells within a module are shaded, they cannot produce current and will become reverse biased. This means the shaded cells will dissipate power as heat. Over time failure of the cell (or cells) will occur. The worst case is a wet maple leaf stuck on the front of a solar panel directly on top of one of the cells. I've seen the shaded cell of a solar panel get so hot that it melted through the back pane of the module.

To prevent this, a *bypass* diode (not to be confused with a *blocking* diode is placed across each module. This diode provides an alternative current path in case of shading. If all

Here is an example of a solar dual-axis tracker.

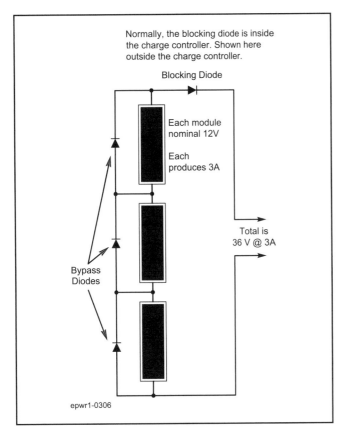

Normally, the blocking diode is inside the charge controller. Shown here outside the charge controller.

Blocking Diode

Each module nominal 12V

Each produces 3A

Total is 36 V @ 3A

Bypass Diodes

epwr1-0306

Figure 3.6—Bypass diodes are added to strings of solar panels. For clarity, only three modules are shown. Normally, bypass diodes are only used on 24 V and higher systems. With the three panels in series, we raise the system voltage while the current remains the same.

is as it should be, the diodes are reversed biased and life is good. Shade a cell and the bypass diode is allowed to conduct, carrying current around the shaded module. **Figure 3.6** shows a three-module series system with bypass diodes added. Normally, bypass diodes are not used in 12 V systems.

In a 12 V system, there is no need for bypass diodes. In a 24 V system it's a good idea. If you have a 48 V system, bypass diodes are a must!

Today, many module manufactures provide the bypass diodes already installed in the module's junction box. If not, I use silicon diodes rated at 6 A attached to the terminal strip inside the junction box.

OUT IN THE WEATHER

Don't place Plexiglas®, glass or anything else in front of the solar panel. All you're going to do is reduce output power. You don't need to protect your solar panel from the rain, snow or weather. Modern solar panels all pass the requirement for being weatherproof. This is known as *block VI testing*.

Don't use reflectors, mirrors or lenses to increase the amount of light striking the panels. This will cause the cells to overheat and actually reduce the amount of power they produce. It's a good chance you will toast the cells beyond their design limits and end up destroying the entire panel.

TILT ANGLE FOR YOUR PANELS

In the northern hemisphere, the preferred azimuth for solar arrays is true south. To optimize performance from your solar panels, the rule of thumb is to tilt the module toward the Sun at an angle equal to your latitude plus 15°. This will give you the best power for the winter, the time of the year when the Sun is spending most it its time behind clouds. To get the most power for the summer months, the figure is latitude minus 15°. This angle of incidence makes

Since my system is wired for 48 V, the panels are wired in series. The bypass diodes are located in the solar panel junction boxes.

This is my current solar array. This array will run about 90% of my house. Under standard test conditions, the array will produce 2.2 kW of power at a nominal 48 V dc. The array consists of twentyfour Solarex MSX-64 modules and eight Astro Power AP-120 modules.

Notice the dual motors used to turn the array east and west while another allows the array to tilt north and south.

sure the light rays strike the surface of the module with a line perpendicular to the Sun.

There's no need to drag out the protractor, plumb bob and string. A simple and quick way of setting the tilt angle requires just a 2 × 4 about a foot long.

First — and this is important — make sure the end of the 2 × 4 is cut as square and perpendicular as possible. Use a power miter saw or radial saw. Don't cut it by hand. The end *must* be square and true.

Pick a sunny day and go out around noon. Place the square end of the 2 × 4 on the solar panel. Hold it in place and note the shadow it casts. Now adjust your solar panel, stopping to hold the 2 × 4 and checking the shadow. Keep adjusting the tilt and position until no shadow is visible. Lock down the hardware and you're done!

Here is a general rule to get you in the ballpark:

Latitude of Site	Tilt Angle
0° to 4°	10°
5 to 20°	Latitude plus 5°
21 to 45°	Latitude plus 10°
45 to 65°	Latitude plus 15°
65 to 75°	80°

If loads are heavier during the early afternoon and evening, some arrays have a west-of-south skew to increase production for the peak load-current demands. This is common when the solar panels are used mainly in a grid intertie system.

In the real world, a solar panel may produce up to 1.5 times its short circuit current due to *edge of cloud effect*. This short-term cloud focusing acts like a lens to concentrate light on the panels. It's possible to see irradiance levels as high as 1500 W/m^2 when puffy clouds are present to focus the sunshine. These edge of cloud effects seldom last for more than several seconds. Few of use will ever see edge of cloud effects occurring.

It's not at all unusual to see solar irradiance at ground level in the southwestern United States exceed 1000 W/m^2. Get on a mountaintop and solar irradiance routinely average 1200 W/m^2.

Of course, not all of us live in the southwest with all that sunshine. In Ohio, we have winters. And some of them can send the mercury plunging. As I have said, heat will reduce the output of our PV module, but cooling the module will increase its output.

On a cold bright January day in Ohio, with the thermometer trying to get above zero, we can easily produce more power than the photovoltaic module is designed to make. You won't see 100 W come from a 20 W module, but the increase can be impressive.

SHORT CIRCUITS AND SOLAR PANELS

The solar panel has another unique ability: You can short circuit the panel without causing damage. Take a look at **Figure 3.7**. As the solar panel is shorted, it will produce its maximum current. That's all. If you were to short circuit the wall socket, the current would try to go infinitely high, limited only by the ability of wire to handle the current before melting.

SOLAR TRACKERS

Generally, when charging batteries with solar panels, I don't recommend tracking the Sun. In my location in Ohio, we sit too far north to get much benefit from tracking. If you're pumping water in Texas, then a dual-axis tracker can increase total power output by as much as 70%! An example of a dual axis tracker can be seen in the photos on page 3-12. A detailed look at the two screwdriver motors can be seen in as well.

KEEP THE PANELS CLEAN

You would not believe the amount of junk that ends up on the front of a solar panel. Keep the glass clean. Bird drop-

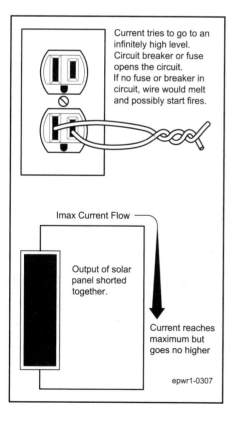

Current tries to go to an infinitely high level. Circuit breaker or fuse opens the circuit. If no fuse or breaker in circuit, wire would melt and possibly start fires.

Imax Current Flow

Output of solar panel shorted together.

Current reaches maximum but goes no higher

epwr1-0307

Figure 3.7— As the solar panel is shorted, it will produce it's maximum current. That's all. If you were to short circuit the wall socket, the current would try to go infinitely high, limited only by fuse or circuit breaker, or the ability of the wire to handle the current before melting.

Here's an example of a simple homebrew solar panel tester. It's just a 200 mV digital panel meter and an 8 Ω resistor. When set to read voltage, the internal scaling resistors set the meter to read 200 V. When set to read current, the meter reads the voltage drop across the 8 Ω resistor.

A combiner junction box with the cover removed. Notice the large wires on the one end and the large number of smaller wires on the other end.

What looks like a TO-3 transistor is really a bypass diode. In this case improper assembly of the interconnecting wiring allowed water to enter the junction box.

Hiding inside this junction box is a bypass diode.

pings can be a problem, so don't mount your solar panels under a beam antenna. The birds will sit up on the beam and poop over your panels all day. Been there, done that.

Speaking of beams, unless you have a small solar panel and a heavy duty tower, it's best not to mount the panel on the tower. During high winds the surface area of the solar panel will produce extreme twisting forces on the tower.

The solar panels must face the Sun. No glow, no go! If you decide to mount them on the southern exposure roof of your home, you can count on an electrical inspector requiring a ground-fault interrupter (GFI) that is rated for dc for use with solar modules.

WIRING PANELS FOR MORE CURRENT: USE A COMBINER BOX

When adding panels in parallel, the current produced by some 12 V systems can become huge. It's not uncommon to have charging currents of over 120 A! Large gauge wire is a must to reduce I^2R losses. A combiner box may be required to simplify connecting all the panels together. A combiner box is nothing more than a large terminal block that takes several smaller wires and combines them into one larger cable. You will need two — one for the positive and one for the negative wires. Generally, this combiner box is mounted out by the solar panels and the heavy cable runs in to the battery room with the charge controller and other balance-of-system components.

It's very easy to combine solar panels. The only thing you should do is make sure you add like-wattage panels together. If you have a 75 W module, then you can combine any number of 75 W modules in either series or parallel. The closer you match the panels in wattage the better the match. Sure, you can add a 50 W module to the 75 W module you already have, but it's better to match

This is but a single solar cell. The ICOM TH7 transceiver is dwarfed by this cell.

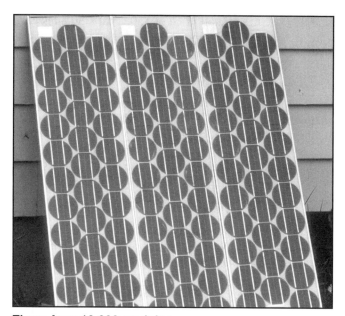

Three Arco 16-200 modules.

wattage for wattage. Remember, you add panels in series to increase voltage and add panels in parallel to increase current.

WHAT'S OUT IN THE MARKET TODAY?

We have a number of choices when it comes to selecting a solar module. There are several players that have been around for a long time. The companies may have changed names, but they still produce first-class products.

Who makes the best solar module? I'll tell you, when you tell me who makes the best HF rig! No one makes the perfect HF rig and no one makes the perfect PV module either. You and I both have our experiences with different rigs and modules. I have my favorites and those are the ones that I lean toward when asked who makes the best PV module.

My first commercial PV module was a 16-2000 made by Arco Solar. I paid a whopping $600 for that 32 W module in 1982. In my solar array that runs my shack and house my array consists of MSX64 modules by Solarcx and 120 W modules by Astro Power.

Arco Solar was purchased by Siemens several years ago. They dropped some models and came out with newer ones. Shell purchased Siemens Solar a few years ago. The new company, Shell Solar, continues to produce quality PV modules. Their SM110 panels run about $525 each. Get two or more and you can find them for around $479 each.

Astro Power ran into some hard times during the later half of 2003. During the second half of 2004, General Electric purchased Astro Power.

Astro Power made a name for itself by making inexpensive silicon wafers. They also made the largest single PV cells in the world. As a matter of fact, Astro Power first started out just making PV cells and selling them to others that made modules.

When you're talking amorphous cell technology, the only player I would recommend is Unisolar. Their third-generation thin-film PVs are outstanding.

The Solarex company was purchased by BP Solar. The new name, BP/Solarex also continues to produce first-class PV modules.

No matter who you pick to supply you with a PV module, you have to remember one thing: You're buying watts. And a watt is a watt is a watt. Read though all the hype and get the most wattage for the dollar. I am sure you have noticed that I lean somewhat towards the dollar per watt figure. It's the best method of seeing what the real price of the power costs you.

If at all possible, get the highest power panel that you can afford. It's cheaper per watt when you do this. For example, a 5 W solar panel costs $100. A 10 W panel sells for $136. The 5 W PV will cost you $18 per watt. For $36 more you get twice the power, or a cost of $13.60 per watt. Go for a 110 W module at $479 and the cost per watt drops to an impressive $4.35 per watt.

While it is easy enough to add extra modules to your system at a later date, the only time you should consider smaller 5 and 10 W modules would be because of size or weight restrictions.

Solar cells come in all sorts of shapes and sizes.

Solar panel manufacturers have a bad habit of dropping models and coming out with bigger and better ones. That's good for the long run, but it can make expanding your system a pain. Although the electrical aspects of module expansion is quite simple (matching the peak voltage), the physical changes can sometimes generate problems. For example, when Siemens Solar produced the popular 50 W SR-50 module, it used a nearly square structure. Siemens later dropped it to make way for a higher-power module. When Shell brought out Siemens Solar, their new 50 W module was *rectangular*. Same power, but different footprint. I guess it is a matter of what you like to see when you drive up to your house. A mixed-and-matched set of solar panels may be the best and cheapest way of getting a system up and running, but it may not set with the neighbors very well.

I would never consider anything under 20 W for a home-based backup system. While low-power (QRP) operators may find that a 5 W panel provides more than enough power for their stations, most of us will need a lot more.

Solar Panels for Portable Use

Like most QRP operators, I enjoy sticking one of the new micro-sized HF rigs into a backpack and heading out in the field. A Yaesu FT-817 or ICOM 703 are great little rigs for taking along on a bike trip. There's a lot of activity on the HFpack frequency of 18.1575 MHz.

Needless to say, anything you need to carry around must be as light as possible. For this application, I like the BP/Solarex MSX10Lite series. This 10 W module uses polycrystalline cells and produces enough power to really do something. There's no glass used so there's nothing to break. The cells are sandwiched between two layers of Tefzel material. The entire panel is then mounted on a stainless steel back frame. While you can't bend the MSX10Lite, it will flex a bit if you drop it. The MSX-10 is about the size of an open magazine.

The BP/ Solarex Lite PV modules come in 5 W, 10 W, 22 W and 32 W. The last two are a bit too large to carry around on a bike.

Really Small Panels

I have some 1 W (yes, 1 W!) modules. They are about the size of a CD holder. These modules can be wired for either 6 or 12 V operation. They're great for charging a set of NiCds or other small rechargeable batteries. They do have one drawback: they are not cheap. That ol' dollar-per-watt thing rears its ugly head when you're talking 1 or 2 watt modules. For example, the 1 W

module I use has a retail price of $30. That's 30 bucks per watt! If you need 3 W of power and buy three of these, you may as well get a 10 W module for $30 more.

Military Packs

All the major manufacturers have solar panels designed for the military. These olive drab folding solar arrays are just as slick as can be, but they sell at prices that only the military can afford. The average price for a 30 W folding military grade solar panel is about a thousand bucks. It's too bad. They make great portable power supplies.

Basic Junk Panels

Every now and then you see surplus panels come on the market. They were normally sold as solar garden lights. The lights are long gone, but these modules show up on the surplus market.

The lights they power consist of an LED or two. The panels recharge a set of NiCd batteries held in the light's housing. So, the peak power of the surplus cell will be about 4 V. Combining several of these panels can boost the voltage up to something we can use. As far as current produced, generally the lights required about 30 to 100 mA, so use that figure as a starting point.

Prices vary based on the voltage and current they produce. I've seen some really nice surplus garden-light solar panels for about $4 each. A good place to start looking would be All Electronics (**www.allelectronics.com**). Another place to check is The Electronic Goldmine (**www.goldmine-ele.com**).

Making Your Own Panels

I really have a hard time with this. I know you have seen surplus solar cells in flea markets and hamfests. They are

On the left side you see Sovonic R100 modules while Arco 16-2000 modules reside on the other end.

there for a reason. They don't work! Sure, those cells may produce power, but there is something inherently wrong with them. Did you ever notice that most of the surplus cells you see laying around in boxes are kind of dull looking? A good first-run solar cell has a shiny appearance. This is because the cells have had a anti-reflective coating applied to them. The manufacturer only applies this coating during the last few steps before assembly into a solar panel. So a dull-looking solar cell did not make it through the quality control stages and thus did not get the coating.

So, that surplus cell you have in your hands may be either out of voltage specs (i.e. producing less than the required 0.5 V), or the current produced is way off. In either case, the cell is out of tolerance. Look at it this way: If the cell was not good enough to make a solar panel for Solarex, why do you think it would make a good solar panel for you?

But what about that bargain price? Well again, this is problematic. It takes at least 36 cells to produce useable voltage. If the cells cost you $5 a pop, that's $180 out of the gate. You will need to buy more than the required 36, too. The cells are so thin, you'll end up breaking some between the hamfest and the ride home. So, let's figure on 40 cells. That makes your first purchase $200.

Now that you have the cells in hand, you need a frame. Nope, can't use wood. Two reasons: (1) It's sure not weatherproof, no matter how many coats of sealer you apply and (2) it will expand and contract at a rate unsuitable for the cells. Solarex found this out when they were making their first generation solar panels. They used a fiberboard back pane and a clear silicon sealer to hold the cells. The back pane would expand at a faster rate than the silicon and the interconnection between the cells would break. When this connection would open, the entire panel would stop producing power. The open panel voltage would drop to zero volts.

You could witness the failure by connecting a voltmeter across the panel and allowing it to warm under the Sun. At some point the voltmeter would show zero volts. Then all you'd have to do was take a water hose and cool the panel. After a few minutes of cool water cascading on the panel, the voltage would snap back up.

So, forget about using wood for a back pane. About the only thing that would work within reason is scrap PCB material with the copper etched off of it. Sheets of G-10 fiberglass would be great if you could find it without the copper.

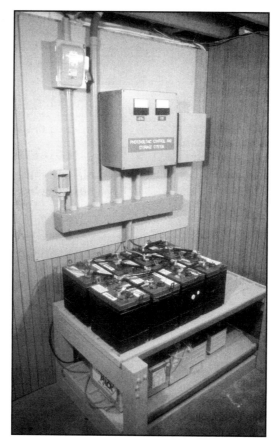

One of my early PV control systems located in the basement, next to the shack.

The next problem would be the covering material. Don't use silicon; you can't normally find it in a perfectly clear form. If you do, it's going to be in the form of a caulking tube. It's almost impossible to apply the material and keep its consistence throughout the entire panel.

Rule out any sort of plastic too. The UV rays will break it down and the result will be a yellow power-robbing covering. Forget window glass, too. It is to fragile and it won't pass all the light to the cells.

So, as you can see, it's not a good idea to try to build your own panel. The money you have invested in cells that may or may not be any good, as well as the material for the frame and covering, outweighs the cost of a commercially made panel — a panel that comes with a 25 year warranty! You can get a brand new 40-W module under full warranty for less than $280.

If you insist on building your own, here are some guildlines that may save you some time and money

- Buy cells that are shiny, not the dull ones.
- Look for cells that have a complete grid pattern on the top.
- If at all possible, get cells with solder tabs already attached. Soldering to a cell that has not been solder plated is impossible.
- Even though the larger cells produce more current, the larger cells are much harder to handle. They will break. Get cells no larger than 3 to 4 inches in diameter.
- Double-sided foam stick tape works best to mount the cells.
- Be mindful of expansion and contraction rates of any of the building material you use.
- For connecting the cells together, use brass shim stock cut into strips with a scissors. Tin the ends with solder and then solder to the cell's solder tabs. Make a small bump between the cells with your shim stock; this will allow some movement if the back pane expands faster than the top cover. The bump supplies the slack to keep the cells from breaking or the brass shim stock from breaking.
- For small homemade solar panels, try using picture frames. They are not perfect, but will produce a somewhat useable solar panel.
- If you must use glass to cover the cells, use tempered safety glass.
- Don't add diodes inside the frame. A failure of a ten-cent diode can be the end of a homebrew panel.

- Don't spend too much of your time on a homebrew solar panel. You can't extract the moisture from within the home made frame, and in a matter of time, all your work will be destroyed anyway.
- Spend your money on a factory made panel and use your time for building the balance-of-system components.

BUYING USED PANELS

Unlike the family Buick, most people just don't wake up in the morning and say, "You know honey, I don't really like the color of those Shell 110 panels. I think we should sell them and get some Astro Power 120s."

Would something like this happen in the real world? Perhaps, but I would have a better chance of hitting the Megalottery and getting a spot on the *Tonight Show* with Jay Leno. Yes, it's possible but the odds are stacked against me.

Do used solar panels show up? Yes, they do. But before you plunk down your dollars, ask about their history. Sometimes the government will scrap a program that used solar panels. Check with the forest extension of a local college for a contact person. A check of the Web for "used solar panels" usually turns up something.

Watch what you pay for them. For the unsuspecting, you may end up paying more for a used panel than for a new one. The guy that is doing the selling may be quoting you a price based on the list price of the panel. This price is almost never used when selling a new panel. The list price can be as much as $300 over the street price. Buyer beware!

If you happen across some used solar panels, make sure you use your multimeter to at least check to see that the panel is producing an open circuit voltage of at least 20 V. You can test for current by using a small value resistor, say about 8 Ω, and measuring the voltage drop across it. A really rough test for current if you don't have a meter handy is to use the same 8 Ω resistor and just feel how hot it gets when the panel is exposed to bright sun.

Avoid first-generation Chornar panels. They were bad when new and you should pass on them used.

Back in the mid '80s Solarex produced a line of solar panels for the forest service and government contractors. Electrically they were the same as their models MSX-64 and MSX-56. The only difference was the OSHA orange frame and back pane. This bright OSHA orange stood out like the proverbial sore thumb! Even if you tried to paint the frame, the orange back pane would give the module away. Now why on earth would Solarex do this? It was their way of trying to keep thieves from stealing solar mod-

ules and selling them at swap meets. The only way you could get an orange panel was to be a government contractor. Even as a Solarex dealer, I was not allowed to order, stock or sell these orange modules. Although Solarex and other companies no longer make this type of orange panel, if you see some be sure you get a bill of sale. You sure don't want the authorities pounding on your door and asking how you acquired those panels!

MOUNTING YOUR PV MODULES

The cost of PV modules can really dent the budget. But no matter what type of module you end up with, you need

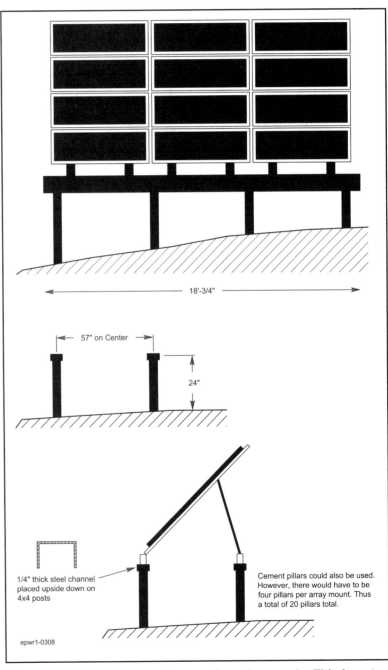

18'-3/4"

57" on Center

24"

1/4" thick steel channel placed upside down on 4x4 posts

Cement pillars could also be used. However, there would have to be four pillars per array mount. Thus a total of 20 pillars total.

epwr1-0308

Figure 3.8—A simple ground mount for solar panels. This is not drawn to scale. Your building requirements may vary from city to city

some place to mount it.

I don't like to measure and drill, then re-drill because I misread the tape, so I buy commercially made PV mounts to hold the panels. Various companies can supply all types of PV mounts from those that mount on the side of a pipe to complete turnkey roof mounting kits. I prefer the ground mounting kits because I like to be able to work on the modules without getting on a roof or tall pipe mount. Plans for a simple ground mount are shown in **Figure 3.8**.

If you're on a tight budget, yet don't want to mess around trying to make your own mounts, then you need to take a trip to your local home center. In the electrical department they sell this stuff called Metal Lumber. It's basically steel angle iron with a zillion holes pressed into it.

The idea is quite simple. You just cut what you need and bolt it together like a large erector set. Save yourself some trouble and be sure to use stainless hardware. Trust me on this. I speak from experience!

Treated lumber placed into the ground serves to hold the frame in place. The whole shebang is easy to assemble and saves you a lot of money. Its money you can use to buy more PV modules!

WIND LOADING

Don't overlook this. It's one thing to set up one or two small modules on a mount, quite another for a large array. Large arrays can have extremely large wind loading. A heavy duty mount is required for large solar arrays. If you

have any doubts, contact a company that installs signs in your local area. They will have the wind loading numbers you need to build a mount. A large array sitting on the ground is like having a dozen 4 × 8 sheets of plywood. In one recent wind storm, I had one Astro Power 120 W module rip right out of the mounting hardware. Build like you're in tornado alley! **Figure 3.9** shows some of the forces the panel mount will have to withstand. The forces placed on the mount will be determined by the wind direction.

Although my homebrew mount has held up rather nicely, I have had two modules blown out from high winds. The only thing holding the panels in place was the conduit and wires!

INSURANCE FOR YOUR SOLAR PANEL

Yes, as odd as it seems, when you start erecting larger solar arrays outside, it's time to contact your insurance agent. You want to have him or her add those onto a rider on your house insurance. It's not that expensive and will make going to sleep at night a lot easier.

SOME COMMON SENSE DESIGN PRACTICES

Most of these are nothing but good old fashioned common sense. In no particular order, here are some of the design practices I have worked with over the years.

Safety First and Last

If you use unsafe equipment or material, it will come back to haunt you. Don't take shortcuts that may endanger yourself or others. Comply with local and national codes. Follow the electrical codes listed under article 690 of the National Electrical Code.

Keep it Simple

Complexity lowers reliability and increases maintenance cost. The more parts in the machine, the more likely those parts will screw up.

Understand System Availability

Don't plan on 99% plus availability with any backup energy systems. Trying to get 100% availability may not be possible or affordable.

Be Realistic When Estimating Your Loads

Figuring in a 35% fudge factor can cost you a lot of money.

Plan On Periodic Maintenance

Solar electric systems are known for their enviable record of unattended operation, but no system will work forever without some maintenance.

Double Check Your Weather Sources

This is very important when doing system sizing. A small error here can set you up for serious performance disappointment.

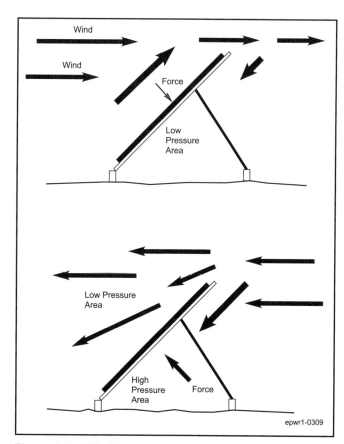

Figure 3.9—Wind forces on a ground-mounted array.

epwr1-0309

Low Voltage Does Not Mean Safe Voltage

Even though we are working with just 12 V, the high currents can be extremely dangerous. Sparks can cause batteries to explode. If nothing else, a sudden spark that comes as a surprise can cause you to jerk, drop tools and cause bodily harm.

Don't Scrimp on the Wiring

Use the right wire for the job. That means the wire must have the ability to safely handle the current and the voltage. If you plan on putting the wire into conduit, then you must derate the amp carrying capacity of the wire.

Use the Correct Rated Fuses and Circuit Breakers

You must use dc rated fuses or circuit breakers when working with dc circuits. A breaker rated for ac use will not function correctly or safely in a dc circuit. The fuses or breakers must have a high amperage interrupt capacity. All dc fuses and breakers should be rated with sufficient AIC to interrupt the highest possible current.

Chapter 4

Charge Controllers for Photovoltaic Systems

Photovoltaic (PV) panels can be added in parallel to increase current to extremely high levels. Since a PV module can also produce operating voltages in excess of the amount we can use, we need a means of controlling the charging current to our storage battery. We need a *charge controller*.

CHARGE CONTROLLERS

In all but the simplest PV systems, some sort of charge controller is necessary. Functioning as an electronic switch, the charge controller protects the battery bank from overcharge and in some cases excessive discharge. Overcharging can boil the electrolyte from your battery bank causing an expensive failure. Needless to say, without a charge controller, you can shorten the operating life of your battery bank. It's also quite possible to damage any low voltage loads connected to the battery without a charge controller. The battery terminal voltage may reach as high as 15 V without some sort of control circuit.

All charge controllers use a little bit of energy; they impress a fraction of loss onto the overall system. This loss is usually so small as to be discounted when dealing with a PV array that produces under 500 W. The controller losses are listed as part of the over all efficiency of the solar electric system.

The cost of a charge controller increases rapidly as the current requirements increase. Controllers that operate at a nominal 12 or 24 V with current to 20 A or 30 A are relatively inexpensive. Controllers that will handle upwards of 50 A are usually 2 to 5 times more expensive. Some controllers are able to handle over 100 A in a single unit. The Heliotrope General CC-120 was able to control 120 A at a nominal 12 V. It had a price tag of about $575. This controller used power MOSFETS as the control devices.

Larger arrays usually require several charge controllers to distribute the current, so no one charge controller must handle high current.

HOW THEY WORK

Nearly all charge controllers work by monitoring the state-of-charge (SOC) of the battery. This SOC changes from day to day and even hour to hour. Nearly all charge controllers use the battery terminal voltage to compute the SOC. While not perfect, the battery terminal voltage does not change all that much during the bulk of charging.

Some high-end commercial charge controllers have the ability to read the number of amp hours that loads have used during the night. At dawn, the charge controller then knows how many amp hours must be replaced in the battery bank. The down side of this type of charging system is that the charge controller can pick up false data from electrical discharges outside the system, such as nearby lighting strikes. The false data can "confuse" the microprocessor.

A well-designed charge

The "Best" Charge Controllers

What's the best type of charge controller to use? There really is no correct answer to this question. By far the most popular type is the series controller using one or more power MOSFETS.

Small solar panels and small capacity batteries in portable applications can get by quite nicely with a small shunt type controller.

Large multi-array installations usually rely on relay-based controllers using mercury displacement relays.

If you want to impress your friends, tell them when the military uses solar panels they prefer the shunt-type charge controller. The people that track such information have found that when a series charge controller fails, the series device (relay, transistor or power MOSFET) will fail with the device open. This stops all charging current from reaching the battery. No battery, no radio, GPS, or laser designators, or any other goodies the military might be using in the field.

But if a shunt type controller fails, the shunt device (transistor, relay or power MOSFET) will normally fail open as well. Since the shunt device can't be turned on, it can't shunt current from the solar panels. This means that charging current will still be routed to the battery. While it is almost a given that the battery will be overcharged, it still is being charged nonetheless. That way, the grunts in the field can still call in fire support, shine those lasers on the bad guy's tanks and more or less continue with the battle.

---

controller can also read the last stage of charging and adjust things according. This occurs when the battery is just about fully charged. It is at this time that allowing an uncontrolled charging current will damage the battery. The charge controller protects the battery under charge by either redirecting the current to another load or another battery. It can also switch the charging current off and then back on.

The SOC can also change because of battery temperatures induced by ambient conditions, or by charging currents. Some charge controllers can read the temperature of the battery (or of the surrounding air) and adjust the SOC to meet the needs of the battery under charge. This temperature compensation works best when the batteries are exposed to extremes in temperature. A repeater site in on a hilltop would be a good example.

A small charge controller is ideal for portable use. This is a shunt type controller with a 2 A capacity.

When the SOC of the battery indicates that it is time to cut back on the charging current, the charge controller will stop the current flow from the PV array to the battery. How this is accomplished depends on the type of controller. As you will see, there are a lot of different ways of controlling the charge current.

DESIGN METHODOLOGY

There are four basic ways a charge controller can control the incoming current from the PV array.

Series controlling. With a series controller, the charge controller opens the circuit between the PV array and the battery. This works just like someone flipping a switch to the OFF position, stopping the charging current in the process. The "switch" can be a relay, bipolar transistor, or power MOSFET.

Shunt controlling. Here, the charge controller turns a switch on or off, but instead of being in series between the PV array and battery the switch is across the PV panel. In a nutshell, the charge controller short-circuits the PV array. I'll explain how that is done in more detail when we look into shunt controllers.

Diversion. In this case, when the charge controller decides that the battery is charged, excess PV current is sent or diverted to another battery or load. Most of the time, the load is a resistive heater used to help reduce the operating cost of domestic hot water.

The last one is known as *array shedding*. Used mainly in large commercial PV arrays, the charge controller function is broken down into sub array controllers. When the battery becomes charged, the sub controllers drop out, removing excessive charging current from the overall system. The sub controllers are really series controllers that open the circuit between the sub arrays and the main battery. This works only when the array is electrically divided and each sub circuit is wired to a separate charge controller.

CHARGING ALGORITHMS

There are several algorithms used by charge controllers. It makes no difference what kind of charge controller is used. Any of these algorithms can be used in series, shunt or diversion type controllers. Even so, you'll see some algorithms such as pulse width modulation used more in series type controllers than in, say, diversion types.

The most basic is the "on-off" type. There are two voltage thresholds or set points. The controller will react based on these set points. The state-of-charge (SOC) is also known as the *disconnect voltage point*. It is at this voltage that the array is disconnected from the battery.

In an on-off charge controller, once the SOC is reached, the charge controller turns off the charging current. When this happens, the battery voltage will immediately drop.

This is a 5 A shunt controller. This controller also supports temperature compensation.

When the voltage drops to the *resume set point*, the controller reconnects the array to the battery under charge. This way, the controller is constantly turning the current on and then off. Most relay-based controllers use this simple on-off charging algorithm.

Some really simple designs delete the resume set point setting. This is by far the least desirable type of charge controller you could use. Once the battery terminal voltage reaches the SOC set point, the relay is opened and all charging stops. Charging does not resume until the battery voltage drops below the circuit's internal hysteresis and restarts the cycle. A good example of this is the float valve in your toilet tank. Once the tank is filled, the water stops. The flow won't start again until the tank is flushed. If the bottom valve seal leaks, once the water leaks down to a certain point, the float valve once again fills the tank. While our batteries don't have valves, they are subject to internal losses. This loss can trip the controller to recharge the battery.

If we were to close the gap between the SOC and the resume voltage, the on time would be greater than the off time. This increases efficiency because there is no charging taking place when the controller is between the two set points in a simple on-off charge controller.

We can exploit this fact by making the SOC and resume set points almost exactly the same. The result is pulse width modulated current being sent to the battery. Since the current is never really turned off, this type of charging algorithm allows maximum charging current to reach the battery. The only trouble with this type of algorithm is that the battery will remain at *state-of-charge* all the time. In other words, if you set the state-of-charge at 14.3 V when the battery is fully charged, it will continue to remain at 14.3 V. The Heliotrope General CC-60 series controllers are afflicted with this shortcoming. This can increase the amount of water used in lead-acid batteries and in some cases shorten their life if they are located in a hot environment.

The third type of algorithm is what is known as *three stage charging*. Here's how it works.

When the battery under charge is low, all the solar panel's current is routed to the battery (just like any other algorithm, all current is routed to the battery when it's discharged) and the battery accepts this as the *bulk charge* mode. After the battery's terminal voltage has risen to the point where the battery begins to reduce the charging current it needs,

the controller switches over to the *absorption charge* setting. Depending on the charge controller, this setting can be maintained by either monitoring the battery's terminal voltage, or simply by noting the elapsed time. Some designs use both. When the battery is nearing full charge, the controller switches over to the *float charge* setting. This keeps the battery fully charged, but at a much lower voltage than the PWM algorithm used by some charge controllers.

Now don't be misled by the marketing people. Even the simplest on-off charge controller acts like a three stage charge controller. Let's look at this a bit closer.

When the simple on-off charge controller is working and when the battery is low, it too will allow all the current to flow from the solar panels to the battery under charge. This once again is the bulk charge. After a few hours (providing you have enough Sun) the battery will start to become charged and the SOC set point will be reached. The controller will turn off, wait and then turn back on. This may only happen once every hour or so. As the battery continues to take the charge, this process repeats and soon the controller will turn off and on more frequently. This sets up the absorption charge cycle. And finally, as the battery nears full charge, the on-off controller cycles very quickly. This thus keeps the battery terminal voltage from staying at 14.3 V all the time. With the rapid switching, the battery voltage will hover lower than the SOC voltage. So, as you can see, even the cheapest on-off charge controller can have the three-stage charging algorithm.

Another algorithm is one that I used on my Micro M+ charge controller project. I call it *time pulse modulation*. In this algorithm the battery under charge is allowed all the current it wants. When the bulk of the charge is complete, the time "on" starts to reduce. As the battery charges, the "on" time gets shorter and shorter, while the "off" time remains the same. During this charge cycle, the absorption charge is being performed and as the battery becomes fully charged, the "on" time is further reduced until it is nothing more than just a quick pulse every four seconds. This keeps the battery fully charged, but without keeping the terminal voltage at the SOC.

There are some who would argue that this type of algorithm is not as efficient as a pure pulse width modulation. While it is true that during those 4 second off periods there is no charging taking place, if the battery is full, it's full. There's an old axiom that states, "If a quart is holding a quart, it's doing the

---

> ## Common Misconceptions About Charge Controllers
>
> There seems to be a lot of misconceptions about charge controllers lurking about.
>
> One of the more common misconceptions is that a charge controller is a regulator. That's not the case. A charge controller does not regulate anything. It's just a switch between the solar panel and the battery.
>
> A charge controller has no "output." This causes concern when someone connects a voltmeter across a charge controller and then notices there is no "output" on the battery connections.
>
> A charge controller won't operate without being connected to storage batteries. If you think your controller has failed, check to be sure your battery is connected and up to snuff.
>
> Another misconception is that a charge controller also limits charging current. No, it doesn't. If your solar panels can produce 15 A and the storage battery is low, then 15 A is what will be sent to the battery. When the battery becomes full, the current will begin to drop. This is not because of anything the charge controller is doing, but rather, the current drops because the battery has reached its SOC.

best it can." Bottom line? You can't put more energy into a battery than what it is designed to hold in the first place.

POWER POINT TRACKERS

The newest thing on the market is the so-called *power point tracker*. Don't be mislead here. We're not talking about solar panel trackers that track the Sun across the sky. These power point trackers are specialized charge controllers.

This charging algorithm changes the operating voltage knee of the solar array. Some power point controllers claim to increase the charging current by as much as 30 percent. While you can't increase the power from the solar array beyond what the panel can produce, you can change the amount of current somewhat by reducing the operating voltage point of the solar array. This increases the current flowing to the battery under charge. So, if your solar panel produces 4 A, then adding a power tracker controller won't give you 20 A, although it *may* increase the 4 A to 4.9 A. The power point tracker works its magic best when the battery is discharged.

One problem with power point controllers is they can really generate a lot of RFI. They're basically switch mode power supplies being run by a solar panel. Another is the cost. An equal capacity charge controller will be cheaper than the same capacity power point tracker controller.

A CLOSER LOOK AT SERIES CONTROLLERS

In a series charge controller, opening the circuit between the solar panels and the battery under charge interrupts the charging current. There are several ways of doing this.
• Bipolar transistor switch
• Relay
• Power MOSFET devices

BiPolar Transistor Switch

The bipolar transistor switch is used when charging currents are generally under 1 A or 2 A. The transistor requires a high base current and this adds to the overall loss of the system. The voltage drop across the collector and emitter of a typical silicon bipolar transistor is about 0.7 V. When trying to control high current, the voltage drop can cause an excessive amount of power to be lost in the transistor. Since power is equal to voltage times current, at 10 A of charging current, a single transistor would drop about 7 W of power. You can add more transistors in parallel to help offset this loss but then you must add equalizing resistors to the emitter leads. All in all, the bipolar transistor series charge controller is a thing of the past.

Relay Series Controllers

As the name implies, a relay is placed in series between the solar panels and the battery under charge.

A relay based charge controller can have extremely low loss across its set of contacts. Unlike a bipolar transistor or even a power MOSFET, the relay's contact may only drop several millivolts at rated current. They are at home when controlling high charging current on the cheap.

As in almost all charge controllers, in a relay based system, the controller's logic will then tell the relay to open when the state-of-charge set point has been reached. Generally, there are two set points. One is the state-of-charge set point. The other is a reconnect set point. Depending on the charging algorithms used by the controller, these set points can vary as much as several volts. The amount of hysteresis between the two can affect the efficiency of the controller.

In a single-stage relay based charge controller, once the state-of-charge set point has been reached, the relay will open. This stops all charging current from reaching the battery under charge. In a single stage system, the solar panel will remain disconnected until the battery voltage drops to a level that the logic deems necessary to reconnect the relay and once again start the charging cycle. The time between the two can be several hours, if not longer. This makes the single-stage series controller very inefficient. And it does not matter if the switch is solid state or a relay. This is known as an on-off type of controller.

To improve efficiency, the resume set point could be moved closer to the state-of-charge set point, but this makes using a relay for the control device unsuitable. The relay would cycle way too fast between on and off states. The answer is to use a solid-state switch. In this case, a power MOSFET is the ideal candidate.

Because there are moving parts in a relay based charge controller, there's always the possibility that the moving parts will fail first. For this reason, most designers shy away from using relays in controllers in out-of-the way locations. Some charge controllers that use relays don't use the old open-framed types; some use mercury displacement relays. This type of relay is completely sealed. Inside there is a

For high current and high voltage dc switching, it's hard to beat the mercury displacement relay. The smaller one is rated at 60 A while the larger unit has a 100 A rating.

reservoir of mercury at the bottom of the relay. When the relay closes, a plunger is pulled down into the mercury. This forces the mercury up and into the electrical contacts. The result is high current switching with reliability. The down side is the relay must be installed vertically.

The relay's coil current can increase operating loss in the controller, too. Depending on the type of relay used, coil current can easily reach 600 mA.

The logic that controls the charge controller must be altered a bit when using a relay. When the relay pulls in, the solar panel and the battery are connected together. Without a blocking diode, the relay will remain "on" all night long. To prevent the unnecessary loss from the relay's coil and from discharging the battery back into the solar panel, special circuits are needed to open the relay at night.

There are two schools of thought here. One is to open the relay's contact and measure the output of the solar panel. If the solar panel's voltage is high enough to charge the battery, the relay is allowed to reclose. If the solar panel is not producing enough voltage, i.e. at dusk, the relay is "told" to remain open. This requires the logic circuit to open the relay's contact, sample the voltage state of the solar panel and then "decide" what to do. Some designers use a time-based algorithm. Every 20 minutes, the controller opens the relay and measures the panel.

Another way is to monitor the charging current. If the solar panel is charging the battery at 5 A, there is obviously no need to open the relay's contact to see if the solar panel is in fact producing current. The problem with this approach is that at certain low charging currents (such as when the array is exposed to streetlights, the relay may never open. The answer is to combine both the current monitor circuits and the timer to eliminate the "relay on at night" problem.

There is one last problem with relay based charge controllers: they're noisy. Not from RFI, but from the clicking on and off of the relay, especially when the battery is charged. Imagine trying to take a catnap in a mobile home with a relay based charge controller snapping on and off all day long.

Although the relay based charge controller does have several inherent problems, there are some advantages, too. The relay contacts exhibit low loss. At high current, you lose very little across the contacts of a good relay. They will also stand up to overcurrent conditions caused by edge-of-cloud effects without damage. They don't generally produce RFI, which makes them ideal for use with radio systems. And, finally, they can handle high voltage and high current easier than even power MOSFETs. This is especially true when dealing with array diversion controllers along with high currents and high voltage. The mercury displacement relays are common with industrial strength solar charge controllers.

The Power MOSFET In A Series Charge Controller

The use of power MOSFETs has revolutionized the way charge controllers work. Instead of a noisy relay, the MOSFET is silent and can handle quite a bit of current.

Modern controller designs are nearly all based on the power MOSFET.

Because the MOSFET is an almost ideal switch, the internal losses of the charge controller are much lower, too. The drive requirements are nil, making the internal loss of the charge controller quite low.

The power MOSFET used in most charge controllers is of the N-channel type. This is by far the most common power MOSFET made. To put the power MOSFET to use, we need to install it in series between the solar panels and the battery under charge. There are two ways of doing that. The simplest way is to place the power MOSFET in series in the negative lead of the solar panel. The controller's logic would then turn the power MOSFET on and off as needed. The problem with this method is you break the common negative between the solar panel and the battery. This "low side" switching works fine for small systems, or for systems that are portable in nature. For larger grid tie or standby systems, switching the negative lead on the solar array is not the best thing to do. All negative leads should be bonded together and grounded to a single ground point.

So, the other place to insert the power MOSFET is to place it between the solar panel and the battery in the positive lead. This is known as "high side" switching. Unfortunately, there is a problem with high side switching using power MOSFETs. To enhance the gate of the power MOSFET, the gate voltage must be at least 10 V higher than the rail it is switching. So, if your nominal 12 V system is fully charged and sitting at 14.3 V, the MOSFET gate voltage must be at least 24.3 V to enhance the power MOSFET. Anything less will force the power MOSFET into its linear region. At high current, this condition would quickly overheat the power MOSFET and destroy it. Rather than operating in a linear state, the power MOSFET in a charge controller must be either completely "on" or "off"—nothing in between.

To produce the required gate voltage, it's customary to use an oscillator of some type, rectifying its output through a voltage doubler circuit and applying the "boost voltage" to the gate of the power MOSFET. But there is a bug in this plan. (Isn't there always?) To generate the boost voltage, the oscillator must produce square waves, and those square waves are full of harmonic energy. So much energy, in fact, that you can easily hear the "charge pump" on your radio. Careful shielding of the oscillator and decoupling capacitors on the solar panel input wires are necessary. To make matters worst, the wires going to the solar panel and the battery can act like antennas, making any noise produced by the charge pump radiate even more.

Because the power MOSFET can be switched on and off at speeds into the megahertz range, we can now control the charging current as fast as we want. This opens the door for pulse width modulation. In fact, today's solid-state charge controllers all use some sort of power MOSFET as a series control device. And while the N-channel power MOSFET is by far the most popular device chosen, it's not the only game in town. There's a relatively new kid on the block and he's called the P-channel power MOSFET.

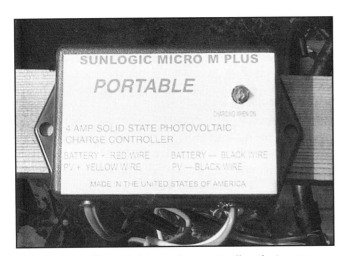

This is a small, portable series controller that uses a P-channel power MOSFET. This controller has a 4 A rating

The difference is the way the P-channel power MOSFET works in a high side application. Instead of requiring a gate voltage that is at least 10 V higher than the rail it is switching, the P-channel can be switched on and off by simply pulling the gate to ground. Or to be more precise, by pulling the gate to the same level as the source lead. In this way, we can drop the charge pump and all the associated filters and decoupling capacitors. The end result is an RFI-quiet charge controller. The P-channel power MOSFET is an ideal device for use with radio communication equipment.

So why don't we see commercial charge controllers using this type of MOSFET? There are two reasons. First, the P-channel MOSFET does not have the same RDS_{on} as its N-channel cousin. RDS_{on} is the resistance between the drain and the source when the MOSFET is fully enhanced. It is measured in ohms. It is possible to find N-channel power MOSFET with an RDS_{on} of as little as 0.009 Ω. While there are some P-channel MOSFETs with that low of RDS_{on}, they are hard to come by.

Why worry about the RDS_{on} of a power MOSFET anyway? Well, depending on the type of device selected, a poorly designed charge controller can use up a lot of valuable solar electric current as heat from in MOSFET.

Here is an example. To keep the math simple, I'll use a charging current of 10 A from the solar panel. We also need to make a few assumptions, like ignoring the overall loss of the charge controller and the wiring to and from the battery and solar panels. We're just looking at the loss from the power MOSFET in this example.

So, back to our 10 A of charging current. We will be using a single N-channel power MOSFET with an RDS_{on} of 0.028 Ω.

We can compute the power by using Ohms Law. Since we know the resistance and the current, we can then compute the power by using the formula: $I^2 \times R$. So, current squared is 100, times the resistance, 0.028. That equals 2.8 W of power being dissipated by the single N-channel power MOSFET. Toasty, but not too bad.

Change the numbers a bit and instead of 0.028 Ω, a more usable P-channel RDS_{on} would be 0.15 Ω. So, once again, current squared is 100, times the resistance, 0.15, which equals 15 W. I have soldering irons that are rated at 15 W and they melt solder! That's hot in anyone's book. And the heat being produced and then thrown away is very expensive because it is ultimately coming from our solar panels.

Now, just for grins, let's run the numbers one more time. This time, we will use an RDS_{on} of 0.009 Ω, a very common values among N-channel power MOSFETs. The power is 100 times 0.009. That equals 0.9 W. See the difference? The N-channel power MOSFET clearly is a better choice when dealing with loss from the series device.

Modern P-channel power MOSFETs have broken the 0.028 Ω RDS_{on} barrier and they are getting lower all the time.

Another reason why we don't see P-channel MOSFETs in common charge controller use is the price. An N-channel MOSFET with an RDS_{on} of 0.028 Ω can be obtained for about two bucks. A P-channel MOSFET with the same RDS_{on} would run about five times that price. While the price has come down in recent years, the P-channel devices are still much more expensive.

THE SHUNT CHARGE CONTROLLER

The shunt charge controller is a strange animal. As the name implies, the shunt controller shunts the charging current away from the battery. It can do this in two ways.

The first way takes advantage of the solar panel. You can short circuit a solar panel all day long without damage. In a perfect world with zero loss in the wire, connectors and shunt device, the solar panel is short circuited when the logic tells the shunt device (usually a power MOSFET) to short circuit the solar panel. At the same time, blocking diodes prevent the shunt device from "seeing" the battery. When the panel is shorted, the current from the solar panel will go to maximum. This is called I_{sc}, (current short circuit) of the panel. But as the panel current goes to maximum, the voltage drops to zero. Since power is based on current times voltage, we have a condition that produces zero power. Why? Because we have no voltage being produced! For example, let's say our solar panel produces 3.3 A I_{sc} under standard test conditions (STC.) With the solar panel short-circuited, the I_{sc} is at maximum, but the voltage produced is zero. Power is current times voltage. So, 3.3 A times zero equals zero watts. This only works with solar panels and nothing else!

But because there is no such thing as a perfect world, and switches, connectors and wire all have some resistance, we *do* produce power when the solar panel is shunted (shorted.) In systems that use a shunt type controller, the entire system must be designed to safely allow for the entire I_{sc} current to flow.

I designed a simple shunt charge controller that would handle up to 5 A of current. It used a power MOSFET as the shunt device. This controller worked just fine for most people—except for hams. They always seemed to connect this controller to a power supply in place of the solar panel. When the battery became fully charge and the power MOSFET short-circuited the power supply, the smoke flew!

A very simple shunt controller can be made with just a 5 or 10 W zener diode of 14.1 V placed across the battery terminals. When the battery becomes charged, the zener diode will start to conduct, shunting away the excess power from the solar panel.

Some shunt charge controllers go an extra step. Instead of just shorting out the solar panel, they drop the excess power into a bank of high wattage resistors. The resistors then dissipate the power as heat. You can take this idea further. If you place those resistors into a water heater, you can use the excess power to help heat the water. In fact, this is what we call a *diversion controller*. The excess power is diverted to another load, another battery or just disposed of as heat.

DIVERSION CHARGE CONTROLLERS

On paper, they work great. In real life, I have found that I normally use more power from my battery bank than I have to recharge the battery bank. So, if there is any excess power to be diverted by the controller, it's very small. The time, work and labor needed to install a diversion shunt controller to help heat my water is a waste. I'll leave the radio on a bit longer and let the battery get fully charged.

Some diversion charge controllers can be used with other sources of power generation. They can be used with some micro hydroelectric systems and some wind turbines, too.

ARRAY SHEDDING CHARGE CONTROLLERS

This is the last type you may see, and normally you won't find this type of charge controller in a ham shack. It's more at home controlling acres of solar panels.

The array shedding controller works by disconnecting sections of the array from the batteries. When the battery bank is discharged, the entire array is directed toward the battery bank. As the battery becomes charged, sub arrays are disconnected from the battery bank. As the battery becomes fully charged, all sub arrays are disconnected while a single series charge controller controls the last one. I use the term "battery" loosely here. In fact, the battery may be the size of a small house and have a capacity of thousands of amp hours.

Each sub array is electrically separated from every other sub set. When I was at the Solarex factory, they had a diversion charge controller running on one of their systems. It used mercury displacement relays. It was slick watching and hearing the relays open and close as the clouds darted in and out.

THE BLOCKING DIODE

Every charge controller should have some method of disconnecting the solar panels from the battery at night. If left connected, the battery can discharge into the solar panels. How much discharge? It varies from cell to cell and module to module, but it is enough to be addressed here, especially when you consider that every bit of stored power is precious.

The oldest and simplest method is to put a diode in series between the solar panel and the battery. During the day, the diode conducts and passes current from the solar panel though the charge control device (say, a MOSFET), then to the battery. At night, the diode prevents the battery from discharging into the solar panel.

There's a price for this simple approach. Even the best Schottky diode has a forward voltage drop of about 0.5 V. At currents up to 10 A, it's not much of a problem. But when you start talking 30 A or more, the loss across the diode can be impressive. At 30 A charging current and a forward voltage drop of 0.5 V, the diode will drop 15 W ($P = I \times E$.) That's a lot of heat.

Although the blocking diode can throw away a lot expensive power as heat, it does provide some safety to the system. If you connect your solar panels backwards, the diode will prevent damage to the charge controller and battery.

As a rule of thumb, solar panels under 5 W don't need blocking diodes. The loss across the diode during the day is more than the loss at night without one. And as I mentioned early on about the relay based charge controllers, they have special circuits in them to prevent the relay from closing at night, thus eliminating the need for a blocking diode altogether. Some smaller solar panels in the range of 1 to 10 W have blocking diodes built in.

When using power MOSFETs in a series type controller, you can put two MOSFETs back to back. In other words, one source lead of one MOSFET is connected to the solar panel. The drain of that MOSFET is connected to the drain of a second MOSFET. This second MOSFET's source is connected to the battery. By turning both MOSFETs on and off, you have made a bi-directional switch. Because the MOSFETs have internal diodes, the pair will still conduct at night. The fix is the same as in a relay base controller. Look at the solar panel and determine if it is producing enough voltage to charge the battery.

I once built a charge controller using 14 N-channel power MOSFETs each having an RDS_{on} of 0.009 Ω. There were seven MOSFETs on each side of the switch. The most I can generate from my nominal 12 V solar array is 57 A. The controller only dropped 5.4 W of power at that current level. Not bad at all!

BELLS AND WHISTLES IN A CHARGE CONTROLLER

Sometimes it is best to keep things simple. A single LED to denote charging is really all one needs. On the other end of the scale, having the charging current, battery and solar panel voltage displayed on an LCD meter is great, too. Remember, the more bells and whistles, the more you're going to pay. And it won't make the charge controller work any better.

The metering in most charge controllers is not known

for accuracy. They will get you close, but that's all. If you want an accurate voltage reading, then break out the Fluke meter.

If you plan on using solar power and the location is out of the way, temperature compensation would be an ideal option. The temperature compensation will alter the state-of-charge (SOC) as the battery temperature increases or decreases. The standard is 3 mV per degree of change. Ideally, the battery electrolyte should be measured, but this is almost impossible to do; the interior of a lead-acid battery is not the most ideal place for a temperature sensor. Most of the time, we simply measure the ambient temperature the batteries are operating in. Although not perfect, it works in most applications.

Another feature you may find in some charge controllers is a low battery load disconnect. Here the charge controller monitors the battery terminal voltage, and when the controller decides the battery has discharged to the point where damage may occur to either the loads or the battery, the loads are disconnected. Once the loads have been disconnected, they must remain off until the battery has been recharged to the point where the battery can once again run the loads.

Some of these low voltage disconnects are relay based. Others use power MOSFETS. The relay based disconnects are more tolerant of overloads and accidental short circuits than MOSFET-based disconnects. On the other hand, the MOSFET-based disconnect requires less drive current and, since there is no relay coil to keep energized, there is less system loss.

OTHER FEATURES TO LOOK FOR IN A CHARGE CONTROLLER

The old Heliotrope General CC-120 charge controller would accept large wiring, up to 4-gauge weld cables for the array and battery leads. This is a great feature to have as you increase the size of your system.

Some high-capacity charge controllers use fan cooling. I like this concept because it does not rely on convection cooling (a heat sink by itself). Some charge controllers live in hot locations such as in sheds and other uninsulated buildings. These buildings can become very hot in the summer time (a good example for the need of temperature compensation, too) and a heavily loaded charge controller can easily overheat.

Another option that is worth looking for is the ability to change the state-of-charge (SOC) without digging out test equipment to do so. Most charge controllers will give you the option of selecting one or more SOC set points by either a dip switch, or by moving jumpers.

Most charge controllers will operate at one nominal voltage. In our case, 12 V nominal systems are the rule. If you decide to increase your solar panels in both current and voltage, some charge controllers can be switched to accept higher system voltages. The nominal systems voltages are 12, 24 and 48 V.

It's also nice to have a charge controller that will allow you to do either an automatic or manual equalization to your lead-acid battery bank. I'll explain more out this when we tackle battery storage in another chapter.

If you plan on operating solar power in the field or portable, then the size of the charge controller may become a factor. Also, if you plan on portable use, the charge controller should be at least water resistant.

And last on our wish list is the ability of our charge controller to provide the necessary protection for our battery without generating RFI. What good is a solar powered radio station if you can't hear the other guy over the hash from the charge controller?

Plan For The Future

Solar power has a bad habit of becoming addictive. You start out with a small panel and some batteries, and the next thing you know you have a back yard full of panels and a garage full of batteries. You should buy or build your charge controller to allow room for expansion. If you purchase a charge controller that has a current capacity of 10 A and your solar panels already produce that much, you won't have any headroom left over when you add on those extra panels down the road.

When a solar charge controller is rated for current, the designer usually takes into account edge-of-cloud effects and derates the controller slightly. If the controller is rated at 10 A, then it should be able to take 20 percent more current for several minutes without problems. This allows for the extra current produced by edge-of-cloud effects and higher-than-1000 W/m^2 irradiance at 25°C.

SIZING YOUR SOLAR CHARGE CONTROLLER

There is a fine balance between the amount of amp hour capacity of your battery bank and the size of the solar panels. If you try to connect a small solar panel to a large capacity battery, you'll more than likely never see the "charged" LED come on. Why? Because the solar panel is only able to produce 200 or 500 mA. The 1200 amp hour battery sees this as nothing more than a float charge—there's not enough current to charge the battery and raise its terminal voltage to the SOC set point. So, the charge controller just sits there, basically doing nothing.

Look at it this way. Let's say that we are using a teaspoon to fill a 55-gallon barrel with water. It is going to take forever to get that thing full of water, let alone enough water to overfill it. It's just not going to happen, especially if we take several gallons out of the barrel every now and then. In the case of a 5 W solar panel and 1200 amp hour battery, there would be no need for a charge controller.

All batteries self discharge. It's the nature of the beast. The self-discharge current would more than offset any charging coming from our small solar panel.

On the other side of this example, let's say our solar panels are producing 10 A and we now have a 17 amp hour battery. In this case, the charge controller will show that the battery is charged within a few minutes. This is because the heavy inrush of current quickly raises the voltage to the SOC point. The battery can't accept the heavy current flow and essentially "fools" the controller into cycling on and off at the SOC.

Let's use that 55-gallon barrel analogy again. This time the barrel represents the solar panel with the water being charging current. For a battery, we only have 17 amp hours so we will represent it with a one-gallon milk jug. If we were to pick up the barrel and try to fill up the milk jug by just dumping all the water out, two things would happen. First, you'd have water all over the floor. Second, you would find the milk jug barely full of water.

As you can see, you do need to be a bit cautious when selecting a charge controller, solar panel and battery. Too small a battery and too much charging current can easily fool the charge controller.

But there are times when you can get away without a charge controller. Remember the first example with the small panel and large battery capacity? You don't need a charge controller in this example. The battery will be more than able to absorb the current from that small panel without damage.

Another example is having a large load placed on the battery with a solar panel connected. The best example is at Field Day. Here you may see a 100W transceiver connected to a 100 amp hour battery and a 75 W solar panel. The radio will be drawing about 20 A keydown and the solar panel will be producing 3 to 4 A under ideal conditions. (Since when is Field Day held in ideal conditions?) In this example, the load current is more than the solar panel current and the battery will be sitting there quite happy, even without a charge controller.

However, if we were to install another 75 W solar panel, then we should be thinking of wiring in the charge controller. Or, instead of our 100 W radio, we stick on a 5 W QRP rig. Then we need the charge controller.

Solar charge controllers are just that. They work on solar electric systems. They won't work with an ac operated power supply, windmill or other power source. While it is true that some controllers do have the ability to handle power sources other than a solar panel, these are the exception to the rule.

Unless you have a really small solar panel and a very large capacity battery, plan on installing a charge controller into your emergency power system. They provide battery protection, a convenient way to monitor your battery, and they let you know what the entire system is doing.

Chapter 5

Generators: Gas, Wind and Water

Producing electricity with solar panels is a very attractive way of supplying your ham station with power when the grid is down. They're a non-rotating, non-polluting, absolutely silent source of electricity — provided the sun is shinning. Batteries can provide a way of storing the power produced by the solar panel for use during cloudy weather or at night. But even with battery storage there are times when you need more power than the solar panels and batteries can provide. There may also be times when your battery and solar panels are at home and you need the power at the repeater site. That's when it is time to fire up the generator.

GENERATOR BASICS

The term "generator" has become rather generic. It is often used to describe practically anything that produces power. I've seen small solar panels described as "solar electric generators." Today, everyone seems to narrow the term down to some sort of engine turning an alternator.

Most people give little thought to generators unless something has gone seriously and suddenly awry — like a total loss of power. In that instance, buying decisions are made very quickly. Both Home Depot and Lowes sold out in a manner of hours during the August 14, 2003 Northeast Blackout.

On the other hand, if you're planning on installing an emergency power backup system for some future crisis, you have the time to do it right.

Like so many things in life, the almighty dollar determines how we go about setting up a backup generator. Before you dig out the checkbook, ask yourself these questions:

1) Do I want to purchase a generator that will last a

Here a 6500-W Honda generator that will run most households

lifetime, or one that might be trashed in a few years?

2) How long will I be operating this generator at any given time? A few hours? Every other weekend at the cabin by the lake. Or will it be several hours every week for months at a time?

3) Will I get any payback from my investment over time?

4) Will I be using this generator for portable or stationary use?

5) What quality of electricity do I need? Will I be running sensitive electronics or a water pump?

6) What kind of fuel is most likely to be available most of the time?

As you can see, taking your time and doing some planning can mean the difference between having little or no power and having the right amount of power at the right location at the right time.

There are several rules of thumb that we can put to good use in the evaluation. Naturally, the cheaper the generator, the less likely it will serve a long and productive life. That being said, if you plan on running the generator once or twice a year, then it may make more sense to get a cheaper generator to begin with.

It's hard to put a finger on payback time. If your location is subject to a lot of power outages, then a more robust generator may save you some money in the long run by supplying power to your house and station.

An important fact to consider is how the generator will be used. If you plan on running the 60-cycle station at field day, then the generator had better be portable. There are times when the power fails in a nearby county and you and your generator are

The Honda 6500 next to the Honda EU1000 inverter generator.

purchase a backup generator is the type of fuel to use. There are four types of fuel used to power the engine that drives the alternator:

- Gasoline. Most common and very easy to obtain
- Diesel. Almost as common and just about as easy to obtain
- Natural gas
- Propane gas

Each fuel type has its own advantages and disadvantages.

The most common fuel, gasoline, can be purchased just about anywhere, and is easily transported. Gasoline should only be transported in approved gasoline cans. A gallon of gasoline produces around 110,000 BTU of energy.

Gasoline engines start easier in cold weather than their diesel counterparts. A recoil starter, better known as the old pull rope, is the norm. Electric start gasoline generators appear in product lines as the engine size grows. And lastly, gasoline generators are very easy to come by. All the generators sold at the home centers are powered by gas.

Gasoline does have its drawbacks. Gas is a good fuel because it burns quickly. This makes gasoline very explosive, especially if the gas fumes find the pilot light on your space heater.

Storage of gasoline is iffy at best. The stuff does have a shelf life. And while you can add stabilizers to help keep it from going soft, you cannot store it for long periods of time without it loosing its potency. If you have ever tried to start the snow blower that sat all summer long with left over gas in the tank, you know what I am talking about. Old gasoline left in the carburetor can really gum up the works. Gasoline burned in small engines produces a lot of pollutants, as they lack pollution controls.

During a power outage, you may not be able to purchase gasoline. Without power from the grid the gas stations can't pump the stuff out of their storage tanks. This really hit home during the Northeast Blackout of 2003.

Since gas does not keep forever, and is explosive to boot, it's best not to keep large amounts of gasoline in storage. But, what do you do if you're caught with a short supply and the power is out? There's no clear-cut answer. One solution is to siphon gas from the family Buick. The gas

needed. Again, portability may be an issue you need to address. Small generators below 5 kW can be manhandled, and can be hauled around in the back of an SUV or small truck. Why, I could even get one into the back of my little Chevy Metro with room to spare for extra gas cans.

When you're talking over 5 kW, then you should look into a wheel kit to at least allow you to move the generator without pulling your back out. A 5 kW generator can tip the scales at several hundred pounds.

When reading the specification of a generator, one thing to look for is a "brushless" design. As the name implies, there are no collection brushes to pick up the power being generated by the spinning armature. This makes for less maintenance, there are no brushes to change and the power to the load is much cleaner. This is an important issue if you plan on running sensitive electronic equipment directly from the generator. If you're running an arc welder, then it's not that important. Plan to spend more money on a brushless generator.

And while on the subject of spinning armatures, you will find that generators fall into two classes. The most common are the gasoline powered 3600 RPM generators and 1800 RPM diesel models. It's not rocket science, but an 1800 RPM generator will last longer than a 3600 RPM generator.

Why 3600 RPM? It's a law of physics that dictates that the armature must spin that fast to produce 60 hertz ac power. The 1800-RPM generators have an alternator that is wound to produce the required 60 Hz at a slower speed.

If we were to impose a load on the generator, say keying your HF transceiver, the load would slow the generator down, causing the output frequency to drop. To maintain the speed of 3600 RPM, a mechanical governor is used. As the load increases on the generator, the governor increases the speed of the engine, holding the output as close to 60 Hz as possible.

The last thing we need to talk about when looking to

The Honda EU1000 is not much larger than a gas can.

tanks on cars can hold anywhere from 10 gallons to 20 gallons or more. With gas prices at an all time high, who can afford a full tank, though?

The next best bet is diesel fuel. It's just about as easy to come by today as more and more automobiles are using it as a fuel. It's less explosive than gasoline, making it safer to store than gas. It too is easily transported to the generator site. As with gasoline, diesel fuel will not keep forever. There are special additives to keep the diesel fuel from turning into jell when stored for long periods of time. During cold weather you need to thin down the diesel fuel with kerosene or other additives to allow the fuel to flow. Diesel fuel will turn into Jell-O when it gets really cold.

Starting a cold diesel engine can be problematic at best. The use of "glow plugs" to heat up the engine cylinder helps cold weather starting. Some diesel engines are hard to start by hand either warm or cold. Most have electric starters. The starting battery needs to be charged at all times and well maintained.

Just like gasoline powered generators, when the power goes out and you're low on fuel, you can't rush to the local gas station to get more if they too are in the dark.

An alternative is setting up your generator to operate on natural gas. This fuel packs a 70,000 BTU bang per one gallon equivalent. Since natural gas is always in vapor form, it is easier to get your generator started during cold weather. Natural gas will not gum up a carburetor like gasoline.

Natural gas is clean burning and is plumbed right into many homes and businesses. During a power outage, natural gas will still be coming into your generator. For long operating times in areas where gasoline or diesel are scarce, natural gas is the stuff to use.

The downside of natural gas is you can't readily transport it from one location to another. And while there are kits available to convert a generator from gasoline to natural gas, I think it best to get one made for natural gas.

Finally, if you want to use a natural gas generator in a portable application, there is propane. Propane has almost as much energy per gallon, 92,000 BTU, as gasoline. Your standard run-of-the-mill barbeque tank holds slightly less than five gallons. Propane is an ideal source for running an emergency generator.

Like natural gas, propane won't foul carburetors. Because it will vaporize all the way down to minus 44°F, it's great for starting an engine when it's cold outside.

Propane stores readily and has an almost unlimited storage life. You can store propane in everything from small 20-pound BBQ type tanks to a 100-pound tank, or even larger tanks for long run times. Since you can refill propane tanks without power at most refilling stations, propane is available during power outages.

Of course, no fuel is perfect and propane does have some disadvantages. First, you must *never* store propane tanks in a building. One leaky valve and...*kaboom!*

Although you can transport propane tanks, the limit to what I want to pick up is a 40-pound tank. Anything larger than that and it becomes hard to move about.

Some of you may have a hard time finding propane. Getting a propane tank filled in my neck of the woods is as easy as eating pancakes. If you live in New York City, well, that could be a problem.

As you can see, there's really no perfect fuel for every application or every generator. How you design your emergency generator system will depend on the type of fuel, size of the generator in watts and of course your budget. But there is one instance where you can have your cake and eat it too. I am talking about a *dual* or *tri-fuel* engine.

Just as the name implies, these generators can run from two or three different types of fuel. The normal configuration is gasoline and natural gas. Next, the tri-fuel is gasoline, natural gas and propane. Depending on who makes the engine; you can switch from one source to another with the flip of a lever, or opening and closing a valve. Having a dual or tri-fuel engine allows using the best fuel source at any given time.

So, when shopping for a generator, read the labels and specifications closely. Sometimes the marketing people write the spec sheets instead of the engineers.

You should check how much the generator weighs. This is one example where the quality is directly related to weight. Cheaper generators are wound using aluminum wire. Heavier, and thus more expensive, generators use copper wire — and lots of it.

GENERATOR GOODIES

There will be plenty of bells and whistles to look at when shopping for a generator. Again, marketing plays a vital role here.

Some generators have small analog voltage meters. While these meters are

A classic China diesel generator. This unit is rated at 10 kW.

nice to have, they're not high-quality Fluke designs! Generally, the meters will let you know if you're close to the voltage you need. Don't depend on them when you're using voltage-sensitive equipment, though.

With today's electrical codes and regulations, all portable generator sets have built-in ground-fault interrupter (GFI) protection. All generator sets have circuit breakers as well. Depending on the wattage of a given unit, there may be a switch to select between 120 and 240 V ac. Some generators can produce both at the same time.

How do you get the power being produced by the generator to the various loads? Most of the time the generator will have several 120 V ac duplex outlets mounted on a panel. If your generator has 240 V ac output, it may also have a twist-lock outlet as well.

For portable use, most of the time, a 240 V ac output would go unused. In a stationary application, a hard-wired 240 V ac output would be more practical, as well as improving overall efficiency.

NEW TECHNOLOGY AND A NEW GENERATOR

There's a new kind of generator being sold today. It's called an *inverter generator*. Basically, it's a generator that drives a solid-state inverter. These generators provide a lot of power in a small package. One inverter sold by Honda at the time of this writing was a 3 kW model that runs on gasoline.

Because this type of generator produces ac by switching power MOSFETS on and off at a very high rate, they can produce RFI. I know of some cases where the noise was stronger than some signals on the HF bands. I can't tell you if they are all like this, or if the problem is confined to a few models. It would be in your best interest to try your loads on one of these generators before you purchase it, or at least make sure you have the option of returning the unit is case it is not suitable for ham radio use.

ENGINES

I've spent considerable time talking about the part of the generator that makes electricity — the alternator. A quality alternator will last and last. It's the engine that wears out over time.

To spin the armature, we need an engine. Generally speaking, it takes about one horsepower to produce 500 W of electricity. So, a generator rated at 2 kW will include a 4 horsepower engine. A 10 kW generator would require a 20 horsepower (hp) engine.

I'm not about to recommend the most reliable generator engines, although I have my favorites. Honda engines are very reliable. Briggs and Stratton engines are known around the world. Tecumseh engines sometimes show up in generator sets, too. In some

locations, you may be able to find parts for a Briggs easier than for a Honda engine.

Look for an engine that has a positive pressure overhead valve. This is commonly called an OHV type of engine. Your $220 generator won't have this type of engine, but the $1150 one will. The bottom line on selecting a power plant with an OHV is longer engine life.

An OHV engine features oil pressure lubrication. Every bearing and bearing surface is coated with oil under pressure for even distribution. This is in stark contrast to the splash oiling system used in a typical 3.5 hp lawn mower engine.

The OHV engines also have oil pressure shutdown system along with a low-oil-level shutdown. This guards against failure conditions where you have plenty of oil in the crankcase but have an oil pressure leak elsewhere.

In any event, spend your money on a generator with a quality engine. An engine that will run for a lifetime may cost you three times what a cheap one will, but the higher price buys reliability and peace of mind.

THE PTO GENERATOR

If you see a PTO generator, chances are you'll be visiting a farm. The *power take off* or PTO generator is powered by power borrowed from other mechanical devices (typically vehicles) that offer *PTO shafts* for exactly this purpose. Besides farm tractors, I have seen PTO shafts on some of the larger hobby tractors like Kubota, Cub Cadet and John Deer. PTO-powered generators can be massive and produce huge amounts of electricity. Upwards of 15 kW are possible, provided the engine running the PTO has the horsepower to spin the generator at that power level.

This is the generator end of the China diesel generator. It's belt driven. The small battery to the front of the generator is used to start the unit.

GENERATOR MAINTENANCE

Needless to say, you have to perform maintenance on your generator. Follow the manufacturer's maintenance schedule. Not only must you change the oil in the crankcase when required, you also have to check the alternator brushes, if so equipped.

Another very important item is to schedule an exercise plan for your generator. Just like people, if you don't exercise the generator, there's good chance it won't work when you need it.

Plan on at least a once-per-month schedule to start. Load test the generator; make sure that the transfer switches work and that all the breakers function. Test the GFI if the generator has one. Don't wait till the lights go out to find out if the generator will start!

Along with extra oil for the crankcase, have on hand an extra spark plug or two. Have some starting ether "just in case," and a tool kit nearby, too.

If your generator has an electric start function, make sure the starting battery is fully charged. A small one or two watt solar panel would be an ideal way to keep the battery charged and ready to go.

A WATT IS A WATT IS A WATT

When you begin looking for a backup generator, the first thing that pops into your head is the wattage, quickly followed by thoughts of cost. After all, while there are specific things to do before you plop down your money on a generator, most of us decide based on the bank account balance. The more money we have available, the more generator we can buy. And what we are buying is wattage. No matter what the price of the generator, it breaks down to how many watts the generator will produce and how many dollars we have to spend to buy those watts.

The wattage rating of a generator is not carved into stone. Some manufactures rate the wattage of their generators as *momentary peak power*. This is the amount of power the generator and engine can produce for a short period of time. Even this time period is sometimes hazy. Is it 5000 W for ten minutes or ten seconds?

What you need to look for is the *continuous power rating*. A 5000 W continuous power rating means just that — the generator can produce 5000 W all day or all week long. Put in enough fuel and maintain the oil level and the generator will produce 5000 W for as long as you're likely to need it.

While this book is slanted towards running your ham station on emergency power, you can use your generator to run many other things, so it is particularly important to keep available power in mind.

If you're thinking about operating appliances from your generator, you'd better consult the appliance manuals or take a close look at the nameplates (the plates that indicate power consumption). If you don't enjoy dragging a refrigerator away from the wall to find the rusty, grime-coated nameplate, there is an easier alternative. I use a little device called a "killawatt." It monitors line voltage, line frequency, power factor and, of course, computes the power drawn by the load. With this device, you know precisely what the load draws in watts, and you will be surprised by how much the wattage on the nameplate differs from what the load really consumes. My 13-inch color TV states it consumes 45 W, when in fact it consumes nearly twice that amount.

Another way to monitor your load current is with a clamp-on ammeter. While this type of meter won't show you the power in watts, it will show you the number of amps being drawn by the ac load. While Ohm's Law is a bit different when working with ac, it's going to be close enough for us. It's a simple matter to convert to watts by multiplying the current times the voltage. So, if your loads are drawing 4 A at 120 V, then the load is consuming 480 W.

Knowing this, it's a simple matter of adding up all the stuff you want to run to determine the total wattage you'll need from your generator. I can tell you right now, forget the electric stove and electric dryer. They consume huge amounts of power and would require a very large generator to operate them.

Determining what loads to run can be tricky. You can't run everything unless you plan on dropping a lot of money to do it. I'll talk about load sizing in a later chapter. For right now, let's assume that you want to keep the basics up and running.

When you have to pick and choose, here are some critical loads to plan for:
- The well pump (if you have a well). Your water supply depends on it.
- Gas or oil fired furnaces need power for the blowers and other controls.
- Refrigerator and food freezer, if you have one.
- Something to cook with — a microwave oven or small hot plate.
- Lights.
- Radio equipment.

Notice that I have radio equipment in last place. When it's your turn to be in the dark for a long period of time, the first things you need to consider are your safety and comfort. Radios can wait.

Depending on the wattage of the generator you ultimately purchase, you should be able to run just about any load in any ham shack. This includes 2 kW amplifier and all the assorted computers and radios. Don't forget the lights!

If you recall earlier, I said that the engine must rotate at 3600 RPM to produce 60-Hz ac. And as you apply loads to the generator, the engine must maintain that speed. The *governor* does this.

To understand why the governor is important, consider that when you're using your generator to operate ham radio equipment, that equipment does not apply a steady load to the generator. For example. If you're running CW, the transceiver may be loading the generator by 50 W or so. Then when you key the rig, the load jumps to, say, 300 W. Remember, that's input power to the radio. The generator "sees" this load change and the governor reacts by increase the engine speed.

The governor will try to match the constantly changing load. In fact, you can copy the CW you're transmitting by listening to the generator load and unload with the keying!

The result is lots of wear on the engine and poor output regulation from the generator. Some newer models have electronic load sensing and speed controls that handle the above situation better than the older mechanical governor does.

One way around this problem is to load the generator in such a way that it sees a constant load. This keeps the engine speed constant, too. While it would seem silly to do, I have seen Field Day sites being powered by a 5 kW generator running a 100 W HF transceiver. The operators loaded the generator down by using several 500 W halogen lights. That solved the problem!

GENERATOR EFFICIENCY

Generators operate at maximum efficiency when they are loaded to about 80% of their capacity. It's just a waste of fuel to run a 5 kW generator for just a few lights and a refrigerator.

That being said, households that live off of the grid, miles from any utility lines, only run their generators when they can load them to their maximum capacity. When they must run the generator, they run it for all it is worth.

By far, the best way to get power to small loads for long periods of time is to charge batteries. Those off-the-grid folks run their generators to charge battery banks. We are talking *big* battery chargers here, not the wimpy 8 A that usually comes from the generators for "battery charging."

While the batteries are charging, they can pump water, run the washing machine and any other heavy load. Not only does this increase the life of the generator, it saves fuel. Five gallons of fuel will last someone living off of the grid a very long time. Generator manufacturers rate fuel consumption at half the rated load of the generator.

THE IMPORTANCE OF A TRANSFER SWITCH

I am surprised by how many of us grab a generator at the home center and then have no clue how to get the power from the generator to the loads. The simplest way is to run a heavy-duty extension cord from the generator to a multioutlet power strip. This is a rude and crude way of getting power to the lights and radios. It's not quick and can be a bit on the dangerous side. There are limits to how much current even the largest extension cord can handle. The longer the wire, the less current you can push through an extension cord.

The best way is to install a *transfer switch*. This will allow you operate from the grid and then, with a flip of a switch, transfer your loads to the generator.

Besides transferring power to the loads, the transfer switch disconnects the house from the grid. This is very important! Without a transfer switch, simply plugging the generator into an outlet would result in "back

This is an automatic transfer relay being installed.

feeding" the grid. This would energize the line from your house on down. If the generator has enough output, you could end up powering parts of the neighborhood. This situation is particularly dangerous for a utility person who may be working on a supposedly dead circuit in your area. An improperly connected standby generator has the ability to kill linemen!

So, let me make this as clear as possible: *Under* no *circumstances connect the output of a generator directly into your electrical service panel without the proper transfer switch.*

You might be tempted to think that you can simply throw the main breaker in the service panel, using it as a kind of crude transfer switch. That's true, but what if you forget? What about the times you're not home and the wife or kids fire up the generator and do not disconnect the service panel from the grid? There are way too many chances to kill someone working on the downed line, or cause damage to your own equipment.

The manual transfer switch located at the generator.

Install a UL-Approved Bypass or Transfer Switch

You can also install a *bypass switch*. This works just like a transfer switch, but installs inside your service box. Mechanical interlocks prevent the use of both the main breaker and the generator at the same time. It is a bit less expensive, and some say it allows more flexibility. Personally, I prefer the standard transfer switch.

To put the transfer switch to use, you need to remove the breakers from your main service box — the breakers that power the chosen loads — and place them in a sub-panel box. This sub-breaker box is then connected to the generator via the transfer switch. When the switch is in the normal position, power from the grid flows into the sub panel and powers the loads.

Flip the transfer switch over to the generator and only the loads connected to the sub panel will be powered. Those loads, and the breakers that protect them in the main service panel, will not be powered. Remember to take into account the wattage that your generator can safely produce when selecting the loads you want to power.

Here are some Web links to some of the products I use:
• Square D # QO4-8M50DSGP.
   http://ecatalog.squared.com/pubs/
• **www.interlockkit.com/intro.html**

WIND POWER

Because we are talking about spinning wires around magnets, the use of wind power to produce electricity seems fitting for this chapter.

When you mention wind power, the vision of an old windmill overlooking a Kansas farm springs to mind. It's an icon that's as American as apple pie. In fact, the name of that particular windmill is an "American fan wheel."

The oldest known use of wind power dates back to the ancient Persians. They used the power of the wind to convert from oxen-driven grain mills to wind power. The term "windmill" kind of stuck and just about everyone calls any type of wind machine a windmill. The correct name for a wind machine that produces electricity is a *wind turbine*, But they will always be windmills to me. I use the two terms interchangeably.

Holland is famous for windmills. The Dutch became world leaders in their design and use in the later half of the 17th century. Today, Holland produces a large amount of is electrical needs from the power of the wind.

Here in the United States, the American fan wheel pumped water all across the Great Plains. Today, they're still in use pumping water for cattle. These installations are being phased out in place of solar electric deep-well pumps, however.

TYPES OF WIND TURBINES AND WINDMILLS

The American fan wheel is a multibladed windmill. The numerous fans or blades allow this windmill to produce huge amounts of torque at very low speeds. This is great for pumping water or even milling grain. It's not that great at producing electricity, though.

My small Winco 200-W wind turbine sitting on the top section of my tower.

Wind turbines like to catch the wind, and that means getting up high. This is a light pole that will be used to hold a wind turbine.

Setting the blade's attack angle.

To make electricity from the wind, you need a combination of high speed *and* good torque. This is best achieved using a two or three bladed wind turbine.

The propeller-driven wind turbine of the horizontal axis type has generally been considered the best wind turbine to produce electricity. The propeller, just like that used in a prop-driven airplane, can be either two or three blades. The jury is still out in the wind turbine court as to which one is better, the two or three bladed machine. In the two-blade corner they say the windmill is much easier to balance and thus will have less maintenance problems. Three-blade proponents say they have no problem balancing three blades and that three blades produce more power per square foot of blade surface.

Two-bladed wind turbines produce more power in higher winds and are thus more cost effective in locations that have higher wind speeds. Adding extra blades improves low-end torque at low wind speeds. If you add too many blades, however, they get in the way of each other and actually spoil the air flow from each other in high winds. This makes multibladed windmills like our American fan wheel self-governing in high winds.

In either case, a horizontal axis wind turbine must be pointed into the wind. This can be as simple as a fan tail like our American fan wheel, or as complex as a computer controlled servo system.

No matter how many blades, or how the blades point into the wind, some method of limiting blade speed in high winds is necessary. Generally, most two and three blade wind turbines have ways of dumping the wind from the leading edge of the blade. This can be as simple as an air brake that causes the wind to spoil across the leading edge, to a centrifugal-controlled pitch control on the blades.

My small Winco windmill had an air brake. No, it's not what you think. This air brake consisted of two flat wings held together with strong springs. When enough centrifugal force was generated in high winds it would overcome the

spring tension and the wings would open. This would cause the wind to spoil over the leading edge of the blade. The blade would lose its aerodynamics and slow down.

The old Jacobs windmills used a set of three heavy balls attached to a lever and gear ring. Each blade had its own ball and lever. Once again, centrifugal force was used to control the speed. When the rotor started to overspeed in high winds, the centrifugal force would move the balls out from the center of the rotor. In the process of moving, the gear ring would turn each of the three blades, destroying the aerodynamics of the blade. This would slow down the rotor while keeping everything in one piece. Since the wind constantly changes, the centrifugal force constantly changes the attack angle of the blades, holding the speed steady during high winds.

VERTICAL AXIS WIND TURBINES

Although you won't see too many vertical axis wind turbines in amateur use, they are found in many other applications. There are two distinct styles of vertical axis wind turbines.

The Savonius Rotor

Developed in 1925 by French engineer S. J. Savonius, the Savonius rotor is displaced in half, and one half is mounted to each side of a vertical axis. It looks like a 55-gallon barrel cut in two and mounted on a vertical shaft. In fact, you can make a Savonius rotor out of two 55-gallon barrels.

The Savonius rotor is not as efficient as a comparably sized two or three bladed wind turbine, but the Savonius rotor has the advantage of accepting wind from any direction without the need to keep the rotor pointed into the wind.

The Darieus Rotor

Another French inventor by the name of J. J. M. Darieus gave us the Darieus rotor in 1927. This vertical axis rotor looks like an eggbeater at first glance. The blades have a symmetrical airfoil similar to a helicopter blade. The blades are curved in a shape called troposkein, which is Greek for "turning rope." Like the Savonius rotor, the Darieus rotor will accept wind from any direction.

The Darieus rotor does have a problem starting from a dead stop. Some designs use wind sensors to measure both the speed of the wind and the duration of the wind gusts. If there is enough wind to generate power a computer tells a small electric motor to kick-start the rotor. Once the wind takes over the motor shuts down

You can also add a Savonius to both the top and bottom half of the Darieus rotor. The Savonius will provide the necessary torque to get the Darieus rotor spinning.

IT'S LOCATION, LOCATION, LOCATION

When buying a home, it's location, location, location. It's the same thing in locating a wind turbine. Most sites don't have enough wind power to produce electricity. What you need to do is a site survey. And you'll need an anemometer. Those small weather stations offered by Peet Brothers and Davis all have anemometers included.

You will also need to find average wind speed, direc-

My Winco sitting on a small short tower. No power at this height!

the wind turbine on a tall tower. And you can't just use the old antenna tower you have laying out back, either. The tower must be designed for use with a wind turbine. As a guess, I would not even think of putting a wind turbine on anything less than a 70-foot tower. I had a Winco wind charger mounted on a 30-foot homebrew tower and only one time did I see any power being generated.

The only exception would be to mount one of the small Air 403 turbines on a guyed, schedule-40 water pipe, and then no more than 30 feet. This would be an ideal installation for a Field Day site. You won't get much power out of an Air 403 turbine, but it will produce enough to keep the radios on the air, provided there is enough wind. The bottom line for windmill installation is high and in the clear.

The amount of wind and the size of the blades determine how much power you can produce. The power a wind turbine can generate depends on the wind's speed and the area of the blade. Power increases with the cube of the wind velocity. If the wind velocity increased from 20 to 25 miles per hour, the power output would *double*. The power produced by a wind speed of 10 miles per hour would quadruple when the wind hits 16 miles per hour.

There are limits, of course, to the amount of power any wind turbine can produce. Once the wind approaches 25 to 30 miles per hour the automatic governing controls should activate. Otherwise, the turbine will likely be destroyed.

WHAT KIND OF POWER DOES A WIND TURBINE PRODUCE?

So far I have talked about what makes up a wind turbine. The power they produce varies from model to model. The smaller units like the AIR 403 produce dc from 12 to 48 V, depending on the unit. My old Winco produced a nominal 12 V dc from its series-wound generator. The old Jacobs wind mills produced 32 V dc. (A complete set of appliances were designed around this 32-V system.)

The Winco windmill I owned had a simple charge controller. It was nothing more than a blocking diode! There was no way to control the charging current short of tuning the blades out of the wind.

Today's windmills have a more complicated charge control circuit. And you can't use a charge controller that is designed for use with solar panels! Most series charge control-

tion and duration tables for your area. You can get this data from the Climatic Atlas of the United States. You can also look up the data from the National Renewable Energy Laboratory on the Web at **www.nrel.gov/**. The most up to date data is in the June 2004, DOE release of *The Report to Congress on Analysis of Wind Resource Locations and Transmission Requirement in the Upper Midwest*. Point your browser to: **www.nrel.gov/wind/uppermidwestanalysis.html**. You can download the PDF file and see just how much wind power is available in the Midwest with this newest data.

Other data from the National Renewable Energy Laboratory is at: **www.nrel.gov/wind/about_wind.html**. This web site is full of site data. Be sure you check it out before even thinking of putting a wind turbine in the air.

Armed with the climate data and your anemometer, you can tell if your site will be one that has the potential to produce electricity from the wind. Generally speaking, you need at least a constant 7 mph wind to produce electricity, and 12 mph wind would be even better.

The wind must be steady. You don't want turbulence, as this will cause the windmill to seek back and forth "looking" for the wind. To get this steady wind, you need to be up high and in the clear. A level, open field will produce more power than a hilltop, depending on surrounding terrain. Again, this should be part of the site survey.

To get to this steady wind, you will need to mount

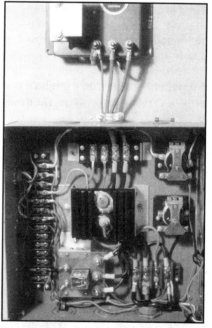

The simple blocking diode used by the Winco wind turbine. It's the small box above the larger box. Inside the large box are more control circuits.

The Winco wind turbine going up.

lers open the current path between the solar panels and the batteries. If you were to connect this type of controller to your windmill, when the controller opens the circuit the windmill would unload, causing it to overspeed. While the controls (air brake, centrifugal pitch) are in place to keep the rotor from overspeeding, the battery bank applies a load to the wind turbine as well. The open circuit would cause the brushes (if so equipped) to arc and burn the armature.

Some load-diversion charge controllers may be used, but here again it is best to use the control system that came with the windmill.

Higher-wattage windmills usually produce ac power that varies both in frequency and voltage. An electronics package then converts the variable power into something that is usable. Most of these systems are normally used in grid intertie systems.

I know of some wind installations that don't produce electricity or pump water. They pump hydrolytic fluid

Setting the torque on each of the bolts to specifications.

A crane is needed to lift this wind turbine onto the tower.

A wind turbine in action.

through a high-pressure pump. This results in heating the fluid to a very high temperature. The resulting heat is used to heat the house!

IS A WIND TURBINE A GOOD SOURCE OF EMERGENCY POWER?

It all comes back to location, location, location. If you think you have a site that is advantageous to wind power, then by all means do a wind survey.

For most of us, however, wind power is just romantic. Wind power gives you a warm fuzzy feeling. Everyone likes

My Winco wind turbine up on its new tower. Even at this height (30 feet) the Winco did not produce enough usable power.

to watch the blades spin. And if those blades happen to be making electricity, so much the better.

MAINTENANCE OF WIND TURBINES

With all those moving parts sitting on top of a tower in the wind, rain and snow, you're going to have to do some maintenance. And if you're not happy climbing tall towers, then either get ready to pay someone to do it, or reconsider installing a wind turbine.

I'm not thrilled with the idea of climbing towers myself. My mind has a built-in altimeter. At about 15 feet it shuts off everything except the desperate need to move in reverse!

HYDROPOWER

Small hydroelectric systems, known as *micro hydroelectric*, are not for everyone. But if you have the water to run one, you will be amazed at the power you can generate. Depending on the water resources, you can figure on generating anywhere from 75 to 250 kWh per month. The amount of power depends on the volume of water and the vertical distance it falls.

A micro hydroelectric system generates power by the force of falling water. You basically take the kinetic energy out of the water the same way you take the kinetic energy out of the wind.

MICRO HYDRO REQUIREMENTS

The number one requirement is water — lots of it. Specifically:
- At least 2 gallons of water per minute of flow and a lot of drop (head)
- At least 2 feet of head and 500 gallons per minute

The amount of water and the drop will determine what kind of generator you can use. There are all kinds on the

The author holds on for dear life at a shattering 15 feet off of the ground. If you don't enjoy climbing towers, forget wind turbines.

market and you should consult someone who is an expert in micro hydro systems to help you decide what kind of generator you need. There are high-flow, low-drop versions and, of course, low-flow and high-drop generators. It's not the generator so much as the design of the water wheel that drives the generator.

You just can't dam up a creek or stream to install your own power station. You'll need to get the proper permits from the local government (read this as lots of red tape). Even if the creek or stream flows through your land, and you have lots of land, you still need the permits.

You may need to run the water from the creek to the powerhouse. The powerhouse is simply a small building or some other structure that houses the water wheel, generator and electronic controls, if needed. If you do, then you need to supply the penstock or pipeline to do so. Just like electrons in a wire, the smaller the penstock the more friction will be applied to the flowing water. So, for longer runs of the penstock, the larger the diameter of pipe. This can get very expensive if you have to run the penstock any distance.

A micro hydro system does have the ability to produce power 24/7. As long as the water flows, you'll have power. Constant power, however, means constant maintenance. You constantly need to check and clean out the water traps to remove logs, leaves and dead critters. Ice and floods can wipe out your system.

There's a whole lot more to discuss about micro hydro

systems, but I'll stop here. Most of us just don't have the water resources to apply a micro hydro system to our emergency power needs, but it is a lot of fun to daydream!

AN IMPERFECT WORLD

For the last several chapters I've talked about generating electricity when the grid is down. We looked at solar energy, wind power, fossil fueled generators and even hydropower. There is not one single source that is perfect.

Solar power is a clean, silent, portable source of backup power. It's not worth a lot if you're site is sitting under a cloudbank. Hurricanes have a nasty habit of making rain for weeks on end.

You just can't count on the power of the wind, either. Just when you need backup power, in comes a high-pressure system and calm skies.

Running a generator for days on end takes a lot of fuel. And when the grid is down, you can't pump the gas out of the storage tanks without electricity.

When designing your emergency power system, plan on using at least two of these methods of producing power. It's clearly a case of one size not fitting all.

Chapter 6

Load Sizing

Perhaps the hardest thing to do when it comes to setting up your own emergency power system is tracking down all the loads. It's amazing how quickly they add up. That becomes even more apparent when you start adding up all the little things you take for granted in your shack.

Putting your station on the air requires enough backup power to get you through until grid power comes back on. How long will you need to be on emergency power? Only you can answer that question. Your result will be determined by your needs and the balance in your checking account.

A SIMPLE EXAMPLE

Since the whole idea of this book is to keep you on the air when the grid goes down, we'll look at a very simple load: an HF transceiver.

When sizing any loads, we need to know two things.
1. The amount of current each load requires
2. How long each load will operate.

Multiply the two and you come up with amp hours. As the name implies, an amp hour is one amp for one hour.

Now at first glance, you may be wondering why we want to use amp hours instead of watts? After all, we talk about operating a 100 W radio. I've never heard of anyone saying they have a 22 A radio on the air.

Well, you could use watt hours but the conversion gets complex. For you see, most of the wattage is based on input power and at 13.8 V. When we compute amp hours, that too is based on input current, but at the nominal system voltage of 12.6 V. So, back to our example.

Let's say we have a typical 100 W HF radio. Pick a model from any company. If the radio is current production and has a microprocessor inside, you can bet the farm that this radio will consume about 1.5 A on receive. Perhaps your model may require less, the other guy's more. For our example, we'll stick with 1.5 A of receive current.

Knowing that, the next question is how long do you plan on running this radio? For the sake of argument, let's say the radio will be used for four hours while the utility company repairs downed lines cause by a storm.

Now we know that the receiver will be on for four hours and it will require 1.5 A. Multiply the time by the current and you end up with the amp hours required for this load. Our calculator shows we need 6 amp hours just to run the receiver.

Now, how long will you be transmitting? And at what power level? Needless to say, the higher the RF output the higher the transmitter's input current will be. As a died-in-the-wool QRP operator, I could tell you just turn down the power and run QRP, but there are times, especially in an emergency, when you need all the RF you got. Less than stellar portable or mobile antennas require more fire in the wire.

It is hard to guess how long the transmitter will be on. Typically, figure a 50% transmit time for rag chew operation and 10% transmit time for emergency communications. In other words, figure the transmitter will be on only 10% of the total receive time.

In the example above, it's easier to convert the four hours into minutes. So, there are 60 minutes in an hour and we plan on four hours so that is 240 minutes. Ten percent of that time would be 24 minutes of actual transmitter run time.

Most of our transmitters run a nominal 100 watts output, and you can figure about 60% efficiency. Doing the math, that means our transmitter will require a dc input of about 20 A. Check your operating manual for the transmitter specifications, or better yet, measure the actual transmitter current.

At the 10% emergency duty cycle, our transmitter will require 8 amp hours. A quicker way to do the calculation is to take one hour of transmit time at 20 A. That's 20 amp hours. Half an hour of transmit time would be 10 amp hours. We need only 24 minutes, so figure a tad less than 10 amp hours.

> Putting your station on the air requires enough backup power to get you through until grid power comes back on. How long will you need to be on emergency power?

With the receiver current and transmit current totaled, we will need at least 14 amp hours. Good design requires that we add a fudge factor to the total. In some circles you can add on as much as 25% to take into account system losses. Let's go with 20%. So, adding in system losses we now need at least 16.8 amp hours.

But all we are running right now is just the transmitter. There are a slew of things that need to be running in our station. One of the most important things to have is good lighting. Keeping the lights on in the shack will require power from our batteries, too.

Now add extras like a two-meter radio, scanner, computer and rotators and you're in for some serious amp hour loading.

And don't forget the little things. You may need power to operate the antenna rotator. Have a remote coax switch? Well then you better add that load to your needs, provided you want to change antennas!

Those niffy automatic antenna tuners require power all the time, too, although some of the newer ones with latching relays do not. I use an SGC 239 and without power applied you don't get anything!

Do you operate CW? Well you had better take the current required to operate any external keyers into account. Yes, I know most newer radios have built-in CW keyers, but some of us still use external keyers that require power. Don't forget PSK-31. You'll not only need power for the interface (if required) but power to operate the computer, too. And PSK-31 will increase your transmit operating time, too.

Don't overlook the phantom loads. Just look behind your table and count how many wall warts are plugged in. During a power outage, all those loads normally powered by the wall warts will need to be on your battery bank. You would be surprised how many amp hours those things require over 24 hours.

As you can see by now, load sizing gets complex rather quickly. By far the hardest part is deciding what to run and when to run it.

DC LOADS ON A BATTERY BANK

Perhaps the hardest thing to do is figure out how long your loads will be on the battery bank. I find it almost impossible to guess how long I will be running any given load at any given time in the future. I have no idea if I will be running HF next weekend or not, let alone for how long. I consider most load sizing as "an estimate based on a guess."

The four hour run time I used in the example above is really kind of weak. In real life how long will an emergency last? Four hours is not much of an emergency in my book. A hurricane can knock out power for weeks on end. Likewise an ice storm can keep you in the dark for weeks or longer. Those amp hours can really add up when you're talking days of operation.

Some loads are easier to size. A repeater, for example, requires that its receiver be on 24 hours a day. The transmit run times vary, but a good repeater site should have some data as to the hours of use on the transmitter. During an emergency, the repeater may be called on all day long,

meaning the transmitter may have an unusually high run time. A good example of this would be a SKYWARN repeater. This repeater would more than likely be keyed down for hours on end as traffic is passed between the spotters and the national weather service. This high usage must be taken into account when sizing a repeater.

When you sit down and start adding up all that you want to run, remember this is an emergency. You don't need to fire up all the stuff in your shack! All you need is the equipment required to carry out communications with another station. That other station can be on 2 meters and across the county or on HF and across the country. Run only what you need to provide emergency communications.

LIGHTEN THE LOAD

Several years ago when the packet network was beginning to grow I had a call from a ham in Colorado. He wanted to operate a remote digipeater from an old oil rig on a side of a mountain. It was a great place for a digipeater. The only problem is there were no power lines to the site. So solar power was the most likely power source.

He planned to use an AEA PK-232 as the digipeater controller. It was given to him by another ham friend, so the cost was low. Okay, free is as low as you can get. But anyway, all he needed were some solar panels, a charge controller, battery and antenna and he would be in business.

The AEA PK-232 required 1.2 A at 12 V. This load would be on 24 hours a day, 365 days a year. It was simple to calculate the amp hours required. It's 1.2 A times 24. That's 28.8 amp hours per day. Add in some system loses and the end result was 35 amp hours per day.

Look at that again. That's 35 amp hours per day just for the packet controller. The radio had to be added in yet. And depending on what his choice was going to be, the radio loads could have been even more. A few quick calculator strokes revealed this set up would need at least three 64 W modules just to run the packet controller.

An old Heathkit HW-202 transceiver was going to be pressed into service as the radio for the network. While the old Heathkit was a good choice, as its receiver standby current is rather small, the run time added up quickly. The 10 W output from the transmitter was really not needed at the site.

The solution was two fold. First, the AEA PK-232 was replaced with a PacComm mini TNC. This thing only required 9 V at 12 mA. The HW-202 was replaced with an ICOM handheld transceiver. He used a high gain antenna to increase the nominal 1 W RF from the handheld transceiver. The entire network digipeater now only required 1.4 amp hours per day. Most of that load came from the receiver!

Since we changed the loads, instead of three 64 W panels, the whole shebang could now be powered with a single 10 W solar panel, with power to spare. Now, instead of almost $1600 worth of solar panels, the entire digipeater could be solar powered for under $150!

The point I am trying to make here is you can really change the load sizing by changing out the stuff you plan on running. Not only can you reduce your battery bank, you can operate similar loads for a much longer time.

SOMEWHAT OLDER RADIOS, LESS POWER

Like the rest of us, I too enjoy all the features that are packed into our modern microprocessor controlled radios. They are also energy hogs. I have no problem with running equipment that does not have a microprocessor inside it.

While I hesitate about recommending an HF radio to use during an emergency, I found that most analog versions of any manufacture work best.

The Ten-Tec Triton Four series of HF transceivers have very low receive standby currents. They all produce 100 W of RF, so you have plenty of power to make contacts outside of the emergency area.

The Ten-Tec Argosy is also a great rig for emergency communications. The receiver draws only 300 to 450 mA, depending on whether you turn off the display. The transmitter can be switched from 50 to 5 W.

In a pinch, an old Kenwood TS-520 is a workable radio. Yes, it does have a tube driver and finals, but you can turn those filaments off when you don't need the transmitter.

Needless to say, during an emergency, especially if running from batteries, don't even think about running any vintage tube-based radios. They require way too much power from your battery bank. You could conceivably run such radios from a generator, but why?

On VHF and UHF, any of the old crystal based radios are good. Yes, you have to have the right crystal pair to work the repeater, but if you're using one of these old rigs from home, you should be good to go.

Radios that work well yet require less receive current are the Heathkit HW-202 and the HW-2036. Look for an old Regency HR-22. An ICOM IC-22S would be nice to have, too.

THE NEW STUFF

You sure can't beat the frequency stability and versatility of today's radios. If you have the energy budget to run them on emergency battery power, then by all means get them on the air!

But you do have to be careful about one small detail. Most of today's microprocessor radios just don't like to see anything less than 13 V. My Ten-Tec Argonaut VI chokes and gags at 12.0 V, well above the low battery discharge voltage of 10.5 V. The same holds true for just about any radio made today. They are designed to run on 13.8 V and just get out and out unhappy with anything less.

The ICOM 703 is one exception that I know of. It will be more than happy with 9 V operation, at QRP power level. The Elecraft company also markets several HF radios that have small energy demands.

Of course this is all a moot point *only* if you plan on running your gear from a nominal 12 V battery system. There are several ways around this problem, but first we need to measure what your loads actually draw from a 12 V supply. Measure current at the voltage at which you plan to operate your gear. We can do that with some simple test equipment.

When we speak about battery storage, we normally use 12.5 V, the voltage of a fully charged lead-acid battery.

MEASURING CURRENT DRAW

While there are several ways to measure how many amps or milliamps each load requires, the simplest method is to insert an ammeter in series with the load. In fact, there are several companies that sell "watt meters" that have nothing to do with RF power! These meters read the current being drawn by the load and then display the results as either watts or amp hours. Of course, some models display more information than others. The higher the price, the more features.

I use two methods depending on what I need to measure. For loads up to 25 A, I use an external shunt and meter. They are calibrated for a full scale reading of 25 A. The shunt has been outfitted with a set of Anderson PowerPoles®. This way, I simple insert the shunt between the load and power supply and read the result on the meter. Yes, a multimeter set to read current will work just about as well, but this is less hassle. For currents under one amp, I use a meter that reads to one amp full scale, without the need for an external shunt. I use this setup for measuring current draw on keyers, low voltage lighting and other low draw loads.

For measuring current over 25 A, I use a clamp-on amp probe. This device measures the current using a Hall Effect device. The results are displayed on an LCD screen. You may have seen these used for measuring ac current. For measuring dc current, the clamp on meter must be rated for use with dc amperage. Most cheap amp probes will not measure dc current. You can get one of these meters that are self contained, or you can get a model that plugs into your existing DVM.

I use the amp probe for measuring ac current. This works best for loads such as 120 V ac lighting as well as battery chargers. Well, in fact, any ac load works with the amp probe.

When measuring ac current with a clamp-on probe, you can only get an accurate reading by clamping on a single wire. You can't clamp onto a line cord and take a valid reading.

The results are shown in ac amps. Although we need to speak in amp hours, when we throw in ac loads, things get messy. As I stated earlier on, we need to speak in amp hours when dealing with battery storage.

If you plan on using a standby generator or a dc to ac inverter, then we need to change our thinking from amp hours back to watt hours.

And the best thing going to measure ac wattage is a little gadget called a "Killawatt." All you need to do is plug in any 120 V ac load and this device will automatically display the wattage the load is using.

So, have I gotten you confused yet? I hope not. Just remember to keep your dc loads in amp hours and the ac loads in watt hours. If you plan on running your equipment from your 12 volt battery bank, then amp hour loading is the easiest way to do things. Running ac loads from a generator or dc to ac inverter makes more sense if you are talking in watt hours.

LOAD SIZING EXAMPLE FOR FUN

Let's do one example so you can get a better feel for how the math works. All you need is paper, pencil and a calculator. If you wanted to, a simple spread sheet with a few calculation fields would make "what if" loading faster.

So, let's plan on running some HF gear, a VHF radio, some lights and a few other smaller loads that I require to keep my station on the air.

I'll be using my Ten-Tec Jupiter for HF and an old and trusty Heathkit HW-2036 FM transceiver for VHF use. Two 12-V fluorescent lights will keep the shack bright. I'll have to have power to the antenna tuner, and I want to keep my small 5 inch TV running. So, time to do some number crunching

On receive the Jupiter requires 1.2 A. On transmit, it requires 18 A. (on a side note, the Killawatt meter tells me on receive, the power supply feeding the Jupiter is drawing 35 W. On transmit it requires 340 W key down into a 50 Ω load. The higher the SWR, the higher the draw on the 12 V power supply and thus a higher demand from the 120 V ac supply. With a 3 to 1 SWR, I have seen the Killawatt meter display over 400 W.)

The lights both require 2.5 A while the tuner current draw is 410 mA. The TV sucks up 800 mA.

Now, for how long? And you know, I don't have a clue. I'll base this example on the August 2003 power outage and say 14 hours.

Okay, the receiver at 1.2 A times 14 hours is 16.8 amp hours. Let's call that 17 amp hours. The next one is tricky. I don't know how long I will be transmitting. I don't know if I will have the only station on the air during this emergency and thus be passing emergency health and welfare traffic, or just telling a control operator on ECARS that Cleveland is out of water. So I'll guess that I will have the transmitter on for 1.5 hours. That's a lot of talking during an emergency.

So once again, take the 1.5 hours times the current needed for the transmitter. In this example, 1.5 times 18 nets us with an additional 27 amp hours. Adds up quickly doesn't it!

Both lights together require 5 amp hours. But when the August 2003 power outage hit, we were under Daylight Saving time, so it did not get dark until four or five hours later that night. Since I have windows in the shack, I don't need to run the emergency lighting all the time. So, I'll figure 5 A for three hours of run time. That's an additional 15 amp hours.

My SGC automatic tuner won't work without power, so it will be on 14 hours. At 410 mA that's another 6 amp hours.

And I liked being able to know that Cleveland did indeed run out of water, so the TV will be on all those 14 hours, too. At 800 mA, that load puts the energy budget in the hole another 11 amp hours. Oops! Let's not forget the VHF station!

On receive the old Heahkit draws 900 mA. Since it too will be on the entire 14 hours that's equal to 13 amp hours. On transmit, the radio requires 4 A. I figure more time transmitting on this radio because the emergency is local as well as national. So I guess about 4 hours of transmit time, or an additional 16 amp hours.

Now, all we need to do is add up the totals for our load requirements.

• On HF we need 27 amp hours.
• On VHF we need 29 amp hours.
• The lighting, tuner and TV require a collective 32 amp hours.

Our station running on emergency power at 12.5 volts will require 88 amp hours. Now, you need to figure in a "fudge factor" of about 20%. Our correct amp hour requirement is 106 amp hours. Whoa! That's a lot of load for only a 14 hour run time.

I have to admit that I loaded (sorry!) this example with a lot of unnecessary loads. The idea is to show you that you can still be on the air and provide a vital source of information to and from your location in times of an emergency with the stuff you have. But to extend your run time — read this as making your battery bank run longer — you can simply change your loads.

I could take the Jupiter off line and put in its place an old Ten-Tec Triton 4. Or put my ICOM 703 on line and run it from a small gel battery. My Ten-Tec Argonaut 5 with its 20 W RF output would more than likely be enough to work the east coast and I would save a ton of amp hours in the process.

Those fluorescent lights could go and be replaced with high efficient LED lighting systems.

While an automatic tuner is fine, and the newer ones don't require power when they are not tuning, I could remove the SGC230 from the antenna systems and break out the old hand crank tuner. The savings would be worth the effort in the long run.

And of course, the run time of the HF and VHF radios could be cut back. The transmit time could be cut back as well.

I hope you can see now how you can trim your loads to make for a more workable battery bank. And that's just what we are going to do next.

SIZING YOUR BATTERY BANK

In the next chapter, I'll discuss lead-acid batteries in great detail. But for now, I'll throw some technobable at you that we need to know to size your battery to your loads.

For this example, we will be using a deep cycle lead-acid battery. We will be discharging this battery to 50% of its capacity. Or, to put it another way, we plan on taking out 50% of the available energy out of the battery.

I have to assume that the battery will be relatively warm, so we will set the temperature derate at "1."

For the batteries, I'll use two very common types. One is the group 27 105 amp hour battery. The second is the T-105 golf cart battery by Trojan battery. This battery is rated at 220 amp hours. This is a 6-volt battery. The group 27 is a 12-volt battery.

The days of storage we will use to compute our battery size has more to do with the account balance in our checkbook than amp hours required. Large battery banks are expensive. The more days of storage, the more money it will cost you.

While most people that live off of the grid say 7 to

10 days of storage is the norm, that makes for a large battery bank. Let's go with 5 days of storage.

Let's get started. We know that the corrected amp hours our load requires is 106 amp hours.

The math goes like this:

Corrected amp hours times the days of storage — in our case 5 days. Divide by the depth of discharge (0.5), then divide by the temperature derate of 1. The result will be the required battery capacity in amp hours.

106 times 5 divide by 0.5 divided by 1 = 1060 This is the required battery capacity in amp hours.

To determine the number of batteries required in parallel, it is a simple matter of taking the required battery capacity and dividing it by the capacity of the battery you plan on using.

If you decide to use the group 27 105 amp hour battery, you would take 1060 divided by 105. You would need ten 105 amp hour batteries to run our loads in the example *for five days.*

Since we are working with a nominal 12 V system, and the battery we choose is 12 volts, the number of batteries in series is one.

Let's try this same example again, but this time we will put in the T-105 battery.

We will use the 1060 required battery capacity divided by 220, the capacity of the T-105. The result is 4.8 batteries. You can't have 0.8 of a battery, so we will round up to 5 batteries.

The T-105 is a 6-V battery, however. To solve for the number of batteries in series, take the system voltage (again we are using a nominal 12 V system) divided by the battery voltage. The result will be batteries in series. With our 12 V system divided by the battery voltage (6) we end up with 2 batteries in series. The total batteries then is the product of the batteries in series times the number of batteries in parallel. In this example, we have two batteries in series times the batteries in parallel. The result is ten T-105 batteries.

Did you notice anything about the batteries? The T-105 has basically twice the capacity of the group 27 battery. Therefore, we only require half as many of the T-105 batteries as the group 27 battery. But since they are rated at only half the voltage, we need to have twice as many of them in series.

If we wanted to, we could specify a J-185 battery. This battery has a rating of 220 amp hours and it comes in a 12 V package. The total required batteries using the J-185 would be just five.

REDUCING THE SIZE OF YOUR BATTERY BANK

There are three ways you can reduce your battery bank. The first one is by far the easiest. Just reduce the number of

things you want to run. Along the same line, you can also reduce the amount of time each load will be running.

The second way is to shorten the number of days of storage. In the example I have given, we specified we wanted five days of battery storage. In some cases, especially with a generator as a backup, we can specify days of storage from two to three, greatly reducing the battery requirements.

The last way is to increase the amp hour capacity of the individual batteries. In a nut shell, it takes a smaller number of larger capacity batteries to run the loads.

There is even a fourth way to reduce your battery bank. While I don't recommend that you do this, it's possible to increase the depth of discharge on your battery bank. In the example above, we set the depth of discharge at 50%. We took out 50% of the energy stored in the battery. You can increase this to 80% with the understanding that such a "deep discharge" will shorten the life of even the best deep discharge batteries. You never, ever discharge a lead-acid battery below 80% of its capacity!

You can also compute the number of amp hours available in your battery bank. It's rather simple. You take the total capacity of your batter bank divide by the depth of discharge you set. In this example, we set the depth of discharge at 50%.

Our battery has the ability to hold 1100 amp hours (using the T-105 in the example and at a 20 hour rate of discharge) At a 50% depth of discharge, we have a usable battery bank capacity of 550 amp hours. I come to that figure by taking the rating of the battery bank, in this case 1100 amp hours times 0.5 (our 50% depth of discharge) which equals 550 amp hours.

If we set the depth of discharge at 80%, then the usable battery capacity increases to 880 amp hours. If we discharge at this depth, we can have a small battery bank at the expense of shorter battery life. I'll explain more about depth of discharge in the next chapter on batteries.

SOLAR PANEL SIZING

Since we are talking about emergency power and battery storage, we may as well talk about sizing your solar panels to your loads.

If you recall, I mentioned in the chapter on solar power that most people size their solar panels not so much based on what they want to run, but rather by the amount in the check book. I'll show you both ways and you'll need to decide what one is best for you and the balance in your checking account.

Sizing Your PV Requirements Based On Loads

In the following example, I use what is known as "back-

> Most people size their solar panels not so much based on what they want to run, but rather by the amount in the check book.

of-the-envelope" sizing. In other words, it is close, but not exact. The calculations are normally made on the back of an old envelope over lunch. I'll discuss some high-end computer based sizing systems later on.

So, to start our back-of-the-envelope sizing, we know what we are planning on running from the example I gave you earlier on in this chapter. We know the loads require 106 amp hours.

We now need to decide a few things. First you need to decide what wattage of solar panel you plan on using. The most common wattage going is the 75 W panel. While there are currently 120 and 150 W modules available, they are near and dear. In fact, you can get 240 W modules if you have Bill Gate's money. For our example, we will settle for the 75 W module.

That 75 W module will produce under standard test conditions (STC) about 4.4 amps. Now we need to know something that you will have to look up. It's the number of hours per day of sunlight at your location. You can get this information from the National Renewable Energy Web page. Just point your Web browser at Google and search.

For this PV sizing example, I'll use my location. The nearest data collection is the Akron/Canton airport. From the National Renewable Energy database, it shows that Akron/Canton Ohio receives 5.6 hours of sunlight in the summer and a dismal 1.2 hours of sunlight in the winter. There is data for every month, but we will use these two numbers for right now.

The math is rather simple. We take the hours of sunlight times the amp rating of the panel. For the summer months, that's 5.6 times 4.4. The result is 25 amp hours (rounded up) for each module.

Since we know the loads require 106 amp hours and each modules generates 25 amp hours, we take the loads divided by the module amp hours. Like so: 106 divided by 25 equals 4.24. Now, you can't have 0.24 of a module, so you can save a buck or two and just say "four" or if you wanted to give a little bit of a fudge factor (and have the money) you could call it "five" modules. Those four modules will run our loads during the summer months, provided we get the average of 5.6 hours of sunlight every day.

If you want to cover your load requirements for the worst case month of January, however, then we take the number of hours of sunlight in January times the 4.4 amps each panel will produce. That's a rather low 5.28 amp hours per panel in January!

To size for the month of January, you once again take the load amp hours divided by 5.28. The result is twenty 75 W modules! Clearly one would need to reduce the amount of loads required during the winter months. This is also a great example of why a hybrid system using a backup generator is such a good idea. The money needed for those extra 15 modules you need to run the loads in January would easily pay for a really, really good generator.

If you look at this result with the 20 modules needed in January, you see why you need that charge controller. In July, all you need is four or five modules, not 20. The charge controller would shut off this array most of the year and your money would be just sitting outside in the sun. Once again, you can see the need for a hybrid system.

Of course, your location will produce different results. If you live in the Sun Belt, then even the January figures will be more doable. In Florida and southern California, you can expect to see more than 5.6 hours of sunlight on any given summer day.

Another Way Of Looking At PV Sizing

While I normally use the above example to size a PV system, you can do it another way and get the same results.

Here, you take the corrected load in amp hours (again 106) divided by the hours of sun per day. In my location that's 5.6 hours in the summer. That results in 18.9 amps that the solar panels must generate to meet the needs of the loads.

When we did the load sizing, we put in a figure to allow for system losses. Some PV sizing guidelines will also add a "fudge factor." It's known as a derated module factor. Normally I use 0.95 as a base line.

So, back to our 18.9 A design current per day. To arrive at the number of modules needed, you take the design current (18.9 A) divided by the module derate factor (0.95) which equals the derated design current. The result is 19.89 A. Let's just call that 20 A. Now, you take the module current at standard test conditions (STC) and divide that into 20 A. That results in 4.54 panels. Once again, you can't have 0.54 of a module, so we round up to five 75 W modules. The result is exactly the same as the first example, although we did add in more of a fudge factor.

Without the fudge factor caused by the module derating of 0.95, doing the math once again leaves us with 4.29 75 W modules.

There's another way of sizing your PV system too. It's in watt hours. But there is an inherent problem when sizing in watt hours. The loads were calculated using the nominal battery voltage of 12.5 V. While the PV modules are rated in watts at 17.1 V. That power point of the solar panels throws things off big time. You can change the amp hours over to watt hours and do the same math, you will end up with slightly different results. Once again, this is because of the power point of the modules being at 17.1 V and not the system voltage being 12.5 V. To avoid pulling out more hairs from your head, just stick with amp hours.

Although we calculated what we needed for PV modules in parallel, I have not shown you how to figure out the number of panels in series. I guess that is because we are working with a nominal 12 V system. Most current production PV modules are rated for a nominal 12 V system, so there is only one panel needed in series. If your system is based on 24 V, then you would need two modules in series. Likewise, a 48 V system would require 4 modules in series. It should be noted that some higher wattage modules only come as nominal 24 V units.

By now, your head should be spinning with all sorts of numbers and figures. What you really need is a way to do the number crunching instantly. That sounds like something a computer would be good at.

Computer Based PV Sizing

With the advent of computers, we now have the ability to play "what if" to change things around in an instant. Having been a dealer in PV modules for a very long time, I have had access to some really slick computer based sizing programs. Allow me to date myself, as some of the early versions were written in *DOS*.

As a matter of fact, I also threw my hat in the ring of computer based sizing programs. While I never released anything to the public, I had a lot of fun coding the software. I can state with utmost certainty I have kept the "goto" spaghetti code alive and well in my coding!

My first attempt was on an Apple IIe using *Applesoft*. I did a smashing of *DOS* and then moved on to Visual Basic for *Windows*. Ohmygosh, I even did one using Apple's Hypercard. I called my creation *SunCad* and it works great, provided you know there are enough bugs in it to keep the Orkin man busy 24/7.

Computer PV sizing allows for instant changes in the type of module selected to the size of the battery. They're really good at doing a cost analysis of the system.

Some programs will also compute the size of fuses, circuit breakers and even what gauge of wire to use. They will also suggest that a hybrid system should be considered when the loads warrant it.

But some PV sizing programs do what is known as "array-to-load ratios." I personally believe this method provides the best results. In array-to-load ratio sizing, you keep a ratio of loads to PV panels from 1.2 to 1.5. An array-to-load ratio of 1 to 1 means there are enough PV modules to run the load, but nothing for battery charging. By increasing the ratio, you ensure that the battery will be recharged and the loads will stay running. Most people using array-to-load ratio sizing stick with the numbers of 1.2 or 1.5 to 1.

In a nut shell, the software needs the climate data for every month of the year. With this information, the software then runs the numbers for each month. It finds the lowest days of sun month and then loops though the sizing math. Each time it loops, it adds on one module until the array to load ratio is met.

Some very high end PV sizing software will also do what is known as "loss-of-load probability." Using extremely complex mathematics the software will give you a probability of when you can expect your loads to run out of power. You can look at it this way: you will know what the up time will be for your system. The loss-of-load probability is given as a percentage. A well designed system will have a loss-of-load probability of more than or equal to 98%. Having 100% up time is neither practical nor cost effective.

Computer based PV sizing programs are available. You will need to so some looking around on the Web. They normally are not free, and some can be out and out expensive. I know of one that does the loss-of-load probability and it goes for over $500 a copy.

On the other side of the road, some of the software is so limited and simple, you may as well use my back-of-the-envelope sizing with a cheap calculator.

If you don't mind a bit of work, you can make a very good sizing program using *Excel*. You just need to add the formulas in a cell and put in the numbers. *Excel* will do the math for you. Granted it won't replace a high end PV sizing software package, but it will do the *what if* planning for you.

About this time you should get up, shake your head a few times and get something cold to drink. All this talk of amp hours and PV current can really have negative effects on you. After you have run the numbers several times, and come up with enough solar panels to run what you want, you come to the conclusion you just don't have the money. This is why I base a lot of sizing not so much on the loads required but by the bottom line in the checking account.

Sizing By Budget

This is the part of system sizing that makes the most sense to me. It goes like this.

You want to be able to run some stuff when the power goes out. You have set aside $500 for the project. That will get you one 75 W module, a charge controller and battery or two.

With that one 75 W module, you take the number of hours of sunlight times the current rating of the panel. In this case we will use 4.4 A.

In summer figure six hours of sun per day. In the winter figure three hours of sun per day. So, just take the 4.4 A times six hours per day in summer and three hours per day in the winter. That's how many amp hours that one panel will produce. Now, take 70% of that figure and you'll end up with a really good real world power per day figure.

Naturally, if you live in a location that has more sun, you'll get more power. Winter will affect the panel's production as the days are shorter. But the figure will really give you an idea of what to expect from your solar panels.

One of the remarkable things about solar is the fact that you can charge up batteries while you're not using your equipment. A cottage by the lake is a perfect example. If you only spend the weekend there, the battery has all week to get recharged. You can get by with much smaller modules compared to what you might normally need.

Sizing For Portable Operation

There are two ways of looking at portable operation. One is to do it just for fun, like camping and hiking. The other one is for operation during an emergency. And then of course there's Field Day operation.

For portable operation just for fun, you need to limit yourself to the amount of weight you can comfortably carry. Especially if you're overweight and out of shape, you don't need to be carrying a lot of radio gear. Then throw in a large glass solar panel and 60 pounds of battery onto your shoulders.

Most camping and hiking trips require QRP radios. Not so much because of their size and weight, but because of the diet-like demands of power.

While I don't like to specify what radios are best for what application, there are several that I know have been real work horses in the field.

The Yaesu FT-817 is super compact. Not only does this radio have all HF capability, but six, two and 440 MHz bands as well. A small foldable two meter beam and the FT-817 would be a lot of fun on a mountain top. The downside to the FT-817 is that it is rather piggish on power. Plan to carry an extra battery and solar panel if you want to spend more than one or two days camping.

The ICOM IC-703 HF radio is quite happy running on battery power. Because it can operate all the way down to 9.3 V, it's ideal for backpacking. But like the 817, it too can load down your battery. The IC-703 plus does have 6 meters, but no other VHF/UHF.

If you don't need SSB, then the Elecraft KX1 Ultra portable CW transceiver is just fantastic. This little guy will run for a week on its own internal battery.

The classic Heathkit HW-8 or HW-9 CW transceivers are very usable. While large by today's standard, they are well suited for portable use. Yup! They will drift when the temperature changes, but they're not that bad. I don't know if I would carry one in my backpack all day, but it has been done more times than you can count.

While the older Ten-Tec Argonaut series can be very happy on battery power, they're just too large to carry into the field. The Argonaut 5 is too battery hungry and gets upset at battery voltages around 12.0 V or so.

The SGC 2020 is the most durable of the bunch. With 20 W of RF, it will put a bit more fire in the wire than the KX1. But at a price — the SGC 2020 is also a bit rough on your battery.

Of course there are plenty of homebrew radios that are just as much at home on the table as sitting on a rock. There are plenty of choices to choose from.

While I am no antenna guru, plan on simple, easy to erect antennas. A small antenna tuner will more than likely be needed as well. Some radios like the IC-703 have internal antenna tuners that work quite nicely without the need to carry an extra box.

If you plan to stay out in the woods for more that two days, and can't plug into the grid, then solar power is a very doable solution to your power needs.

A 10 W BP solar "lite" module is perfect. This panel produces up to 10 W under full sun, that's roughly 580 mA of curent. That is about 3.5 amp hours per day in the summer. This module is about the size of an open magazine. There are four holes, one in each corner, that you can use to bungie cord the panel to your backpack. With the battery and charge controller inside, you can recharge your battery as you walk the trail. There is no glass used in this module, so there's nothing to break. There is no frame either, so the module is very light weight. It is waterproof and can withstand a heck of a lot of punishment.

If you opt for a flexible panel, you better hurry. The only one that I know of is the Unisolar brand. As of this writing, Unisolar has discontinued all panels under 64 watts.

The 10 W solar panel will provide enough power to

Field Day is an exercise in setting up and getting on the air with backup power systems.

operate your gear, is light enough to carry with you and you can use it to power other loads. The panel will easily power most consumer electronics provided you have the necessary voltage converter. The converter is nothing more than a small regulator to convert the 17 V from the solar panel down to whatever you need for your load. Any of the multi-voltage thingies that plug into a cigar lighter in your car will work.

During emergency operation and portable conditions, you may need to tug along a much heavier battery and larger solar panels. Panels up to 32 W are made under the "lite" construction. A typical 75 W solar panel made with glass and aluminum frame is a bit much to carry around.

QRP power levels may not be adequate. Higher power means larger batteries and stronger solar panels. Plan on using VHF and UHF if at all possible.

If your portable setup is in a location you can drive to, then by all means a small generator would be ideal. A Honda EU1000 inverter generator will provide more than enough power for lighting and a 100 W HF station.

As for Field Day, well just about anything goes. Battery, solar, generators or even human powered HF radios. There are an almost unlimited number of possibilities. As it was intended to be, Field Day is an exercise in setting up and getting on the air with backup power systems. Be it a generator or battery power. Antennas, radios and power sources are all tested.

Some of us describe Field Day as nothing but a contest. And when you look at it, you can sometimes see their point. And speaking of contests, if you plan on running emergency power during a contest for the extra points, great! But have you ever given a thought about running Sweepstakes on emergency power? It's a great way to test your gear, battery supply and generator system. Besides, it's a lot of fun! Just like the planned emergency drills, running your station under contest conditions will surely find the problems. Then the next step is to correct them and make whatever adjustments you need. Then when the real one hits, you're ready.

I hope your head is not spinning after all the numbers I have been throwing at you in this chapter. Sizing is an important part of any emergency backup system. It should not be overlooked or stuck on the back burner. That being said, just plan on running what you need when you need it. The best solar electric system is kaput if the emergency hits during a week of cloudy weather.

RUNNING YOUR LOADS ON GENERATOR POWER

The best way to size your ac loads is to add up all the wattage required for each ac load. Then the run time for each load. You'll end up with watt hours. The Killawatt meter will give you an exact wattage of each load.

When we talk about ac loads, then we almost always speak in watts. A toaster is 1200 W. We don't say the toaster uses 10 A. And just like amp hours, a watt hour is one watt for one hour.

Figure on loading down your generator to within 80% of its capacity. Running much less than that is a waste of fuel. If you have a battery bank, now is the time to recharge it when you're running the generator. Run your loads and get the batteries recharged as quickly as you can. Fuel may be hard to obtain if the outage runs for a few days.

Ice storms can last for weeks. Most likely, you're an-tenna will come down with the ice. Generators run out of fuel and no power means no gas from the pumps. I guess it all boils down to the old saying, "The best laid plans of mice and men sometimes go astray."

Knowing what to run and for how long will see you through most electrical outages. If you have the knowledge that your battery bank can provide power to your loads two days you're far better prepared than most. Once the crisis begins, you can adjust what you want to run and for how long. When the grid goes down, nothing is really carved into stone.

Chapter 7

Holding Your Volts: Battery Systems and Storage

You can't mention emergency power and not include batteries in the same sentence. Without a doubt, the use of a battery for electrical storage is a must when using solar and wind power. Batteries play a critical role in emergency power communications. Even if you have a 10 kW generator, you should still use batteries to run your loads.

The problem with batteries is there are so many different types to choose from. Even within the same chemistry family, there are choices to be made.

BATTERY TECHNO BABBLE

Even the way we talk about batteries has become blurred. We talk about getting some "flashlight" batteries at the store, when we want some "D" cells. In reality, those aren't flashlight *batteries* at all. What you get at the store when you buy "D" cells is a package of single cells. You need to connect the cells to make up a battery. The common terminology is to refer to these things as batteries, even if they are single cells, though. For example, the most common "consumer" batteries are "AAA," "AA," "C" and "D" cells. Of course there are also 9-V "transistor" batteries and even 6-V and 12-V "lantern" batteries. That last group really are batteries, consisting of 6, 4 and 8 cells, respectively, inside the metal case.

Batteries don't store electrical energy. They use a chemical process to produce electrical energy. A capacitor can store an electrical charge, a battery can't.

When you discharge a battery, you convert the chemicals inside from one type to another type. Various battery types use different chemicals, and the chemicals produce a reaction that produces electrical energy. The difference is what's inside of the battery.

PRIMARY CELL BATTERIES

Once again, we call these flashlight batteries. They are made for a single use. When they no longer run the load correctly, we toss them in the trash. The old carbon-zinc battery is a classic example of this technology. When the chemicals inside this battery were consumed, the battery was dead. As a matter of fact, those carbon-zinc batteries were so bad, that in the early days of battery chemistry you could not power a flashlight for more than a couple of hours at time. Remember that little push button on the old flashlights? It was there so you could flash the light on when needed. That's where we got the term "flashlight."

Carbon-zinc has been mostly abandoned in favor of the alkaline battery. It's the most popular one-time-use battery sold today. And it too is a primary cell. When it is discharged, it is tossed in the trash.

Several companies tried to make a rechargeable alkaline battery. I had some of them, but after a few charge cycles, the run times were shorter and shorter. They did not last more than a few years on the consumer market.

RECHARGEABLE BATTERIES

While primary cells and batteries were designed for one time use, secondary cells are designed to be used over and over again. In effect, they are rechargeable.

Nickel Cadmium (NiCd) Batteries

Just about everyone has had some device that uses NiCd batteries. They are the main battery type used in consumer electronics devices. They're cheap to produce and have been around for years. My old Drake TR22C used nickel cadmium batteries way back then.

They are easy to recharge and you can get upwards of 1000 cycles before the battery is kaput. They come in most of the common consumer sizes such as "AAA" "AA" "C" and "D" cells. NiCd batteries are easy to come by. You can get them just about any place that sells batteries. I've picked them up at Lowes for a buck a piece in the lighting section.

NiCd batteries have come under fire recently because they contain one of the worlds most toxic chemicals, Cadmium. With millions of worn out NiCd batteries hitting the landfills, they are creating a problem.

NiCd batteries also suffer from "memory effect." In a nut shell here is how the memory effect works. Suppose you use your cordless phone for one hour and then recharge it.

After some months, the battery pack would only work for one hour. It memorized the fact that one hour of discharge would then make for 16 hours of recharge time. Once you set the memory, the cell will not work longer than the memory of discharge and recharge cycles it knows.

The people that make NiCd batteries say their batteries do not have memories. The reason why your cordless phone only works for one hour is the fault of the charging circuit the phone was designed with. Actually, the problem is traceable to the nickel electrode used. This nickel electrode will crystallize, leaving part of the battery unusable.

The best way to avoid this memory effect is to discharge the battery to the point it will no longer operate the load. A heavy discharge and then recharge will erase the memory effect on most NiCd batteries. So, to prevent NiCd memory problems, discharge the battery pack on your hand held to the point where it no longer can operate the hand held. Then recharge the pack.

Because of the toxic nature of the NiCd battery and the so called memory effect, most of us have moved on to the next best chemistry used for rechargeable cells. The Nickel Metal Hydride battery.

Nickel Metal Hydride (NiMH) Batteries

The Nickel Metal Hydride battery is very close to a NiCd from a chemistry point of view. Instead of cadmium they use a form of a metal hydride in the mix. Since the NiMH battery contains a nickel electrode, they too suffer from memory effect. But since the NiMH battery does not have as many discharge and charge cycles as the NiCd battery, we don't seem to notice the memory effect as much with the NiMH battery.

When NiMH batteries first became available, they were very expensive. Today, they are just about as much money as the NiCd battery. They are also available in most common sizes from the "AAA" to the "D" cells. You can purchase a NiMH battery back for most of the common hand helds. They can also be recharged with the same type of charger used by NiCd batteries.

Disadvantages of NiMH Batteries

First and foremost is they don't quite have the same cycle ability as the NiCd battery. Getting over a thousand charge and discharge cycles on a NiCd is easy. You should expect about 200 to 500 cycles with a NiMH battery.

With a higher internal resistance, they can't produce as high a current as either NiCd or alkaline batteries, and thus are not "as powerful" as those types. That said, they do offer a huge increase in amp hour ratings. A standard size "AA" NiCd battery ran from 600 mAh to 900 mAh. Today, the NiMH battery in the "AA" size can have current ratings of 2700 mAh!

They also have a heavy self -discharge. They will loose about 30% of their charge per month just sitting around doing nothing.

While NiMH batteries have a lot going against them when compared to NiCd batteries, they have one thing the NiCd does not — that toxic cadmium inside. With the growing number of consumer products switching over to NiMH

battery chemistry, the NiCd may become our next carbon zinc battery.

And although the NiMH battery is growing in popularity, there's a relative new kid on the block: the Lithium-ion battery.

Lithium-Ion (Li-ion) Batteries

We're seeing this chemistry in some battery packs for our handhelds. The Li-ion battery is proving a superior energy storage chemistry. It too is without fault.

Li-ion batteries come in both primary, one-time-use batteries and secondary rechargeable batteries.

Most of the one-time-use Li-ion batteries show up in cameras and some flashlights. They can produce an amazing amount of current in a given amount of space. My Streamlight Scorpion with its Xenon bulb produces up to 6,500 candlepower with only two 3 V lithium batteries. The only problem is the cost of the batteries — about $9 a set. Li-ion rechargeable batteries are finding their way into consumer goods. My new Apple laptop iBook is powered with a Li-ion battery pack. Ditto my Yaesu VX-6 handheld. Li-ion battery technology is the up and coming chemistry to watch. But it does have its share of problems.

First and foremost is the lithium used in the cells. Lithium is extremely reactive. Lithium and water do not mix. Not peacefully, that is. Bring the two together and you can generate hydrogen and make things explode. To prevent Li-ion batteries from becoming little bombs in your flashlight the lithium is in the form of a lithium salt, which is a much more benign form of the metal.

When you recharge your Li-ion battery pack, you *must* use the correct charger. There are special circuits that monitor the charging process and keep the Li-ion cells happy. Sometimes the circuit that controls the charging is built-into the battery pack. Unlike the NiCd battery, you must use a charger designed just for the Li-ion battery.

The self-discharge rate of the Li-ion battery is quite good. Only about 10 percent discharge per month. So a fully charged Li-ion battery will last on the shelf three months longer than a NiMH battery.

While the Li-ion charge will last longer than the NiMH battery, the Li-ion battery itself won't last as long as the NiMH battery. Figure about two to three year of life from the Li-ion battery, even if you never use the battery!

ENERGY DENSITY

This is a fancy way of stating how much bang you can get in a given weight. The standard is watt-hours per kilogram. If the energy source is a liquid, then the measure is watt-hours per liter. Diesel fuel has 10,700 Wh/l by volume and 12,700 Wh/kg by mass. Since we're talking batteries here, this is how the four compare:

• Li-ion   150 Wh/kg
• NiMH   100 Wh/kg
• NiCd   60 Wh/kg
• Lead-acid   25 Wh/kg

That brings us to my favorite battery, the lead-acid battery.

Lead-Acid Batteries

There is nothing special about this battery. It's been around for a very long time, and the original version from the 19th century has changed very little.

Basically, the flooded cell lead-acid battery consists of two lead plates submerged in a weak acid bath. The sealed lead-acid battery, called a Gel Cell® or Absorbed Glass Mat (AGM) battery is relatively new, coming to the scene in the '70s.

How They Work

I really don't want to get into electrochemistry, because it gets complicated almost immediately after the first two words. So, I'll look at this as simply as possible. Inside the lead-acid battery you will find two lead plates and an acid.

1. Lead dioxide (PbO_2), the material on the positive plate.
2. Sponge lead (Pb), the material on the negative plate.
3. Sulfuric acid (H_2SO_4), the electrolyte.

When two unlike metals such as the positive and the negative plates are immersed in sulfuric acid the battery is activated and a voltage is developed between the two plates. The voltage depends somewhat on the material used in the plates and the electrolyte used. In a typical lead-acid battery, the difference is about 2.1 V per cell. As soon as we connect the positive and negative leads together via a load, a chemical action begins.

Discharging Lead-Acid Batteries

With a load applied across the battery's two plates, current flows and the battery discharges. The lead dioxide in the positive plates is a compound of lead and oxygen. Sulfuric acid is a compound of hydrogen and the sulfate radical. As the battery discharges, lead in the active material of the positive plate combines with the sulfate of the sulfuric acid, forming lead sulfate ($PbSO_4$) in the positive plate.

Battery Repair

When I was a little stinker there was very little money to be had. My old man worked in the local steel mill and there was little left over for car repairs. So, when the battery in the old Buick started to slow down, my old man did what everyone else did. He went to see Paul. Paul was the local fix-it guy where all the farmers took their stuff to get repaired. And if Paul said something, no matter what, someplace on the planet it was carved into stone. So let it be said, so let it be written!

Paul looked the old Exide over and said, "Yup, it needs fixed. Here's what you have to do. Go home and place an aspirin in each one of the cells. That will fix it up as good as new."

Now my old man's logic always amazed me. If Paul said one aspirin, then two would be better and seven about right. Plop! Seven aspirins went into each battery cell.

Of course, that did nothing to fix the battery and in a short while the battery died. To my knowledge, the battery never suffered from headaches, though.

Oxygen in the active material of the positive plate combines with hydrogen from the sulfuric acid to form water in the electrolyte.

All the time, a similar reaction is occurring at the negative plate. Lead of the negative active material combines with sulfate from the sulfuric acid to form lead sulfate in the negative plate.

As the battery discharges, the acid becomes weaker and the two dissimilar plates are becoming more alike. Since the battery plates are becoming more alike, they can no longer supply the voltage required by the load. At some point the voltage will drop to a value below that which will operate the load. The battery is said to be discharged.

A lead-acid battery gives off electric current as the positive electrode's lead dioxide (PbO_2) is converted to lead sulfate ($PbSO_4$) at the same time as the negative electrode's pure lead (Pb) is also converted to lead sulfate ($PbSO_4$). To make this possible you also need an electrolyte around the electrodes, and that is diluted sulfuric acid (H_2SO_4), normally around 30%, which provides the sulfate ion that is needed for the recharge.

When the battery is recharged by applying an electrical current back into the battery, the process is reversed and the sulfate ions go back into the electrolyte. Rest products are heat, hydrogen (H) and oxygen (O_2). The full reaction can be described as:

$$Pb + PbO_2 + 2H + 2HSO_4 => 2PbSO_4 + 2H_2O + 2e-$$

See, I told you it gets complex quickly!

Plates Inside Lead-Acid Batteries

Flooded-cell lead-acid batteries have plates that are mostly pure lead. Pure lead is way too soft for use in a battery, however. So, most companies alloy the lead with antimony, which makes the plates harder. This allows for thinner plates and thus more amperage per cell.

The only problem with antimony is that to the battery it's a slow poison. The antimony will also increase water usage. Nearly all deep cycle lead-acid batteries have antimony alloyed plates.

To help reduce water usage, instead of antimony, you can use a mixture of lead-calcium. The lead-calcium provides a longer shelf life and requires much less water in between charging cycles. Nearly all *no-maintenance* batteries and sealed lead-acid batteries use lead-calcium alloyed plates.

Recharging Lead-Acid Batteries

The chemical actions that take place within a battery during charge are basically the reverse of those that occur during discharge.

The lead sulfate in both plates is split into its original form of lead and sulfate. The water is split into hydrogen and oxygen. As the sulfate leaves the plate it combines with the hydrogen and is restored to sulfuric acid. At the same time, the oxygen combines chemically with the lead of the positive plate to form lead dioxide. The specific gravity of the electrolyte increases during charging because sulfuric acid is being

This photo shows a hydrometer used to check batteries.

formed and is replacing water in the electrolyte.

The battery will produce gas when it is being charged. Hydrogen is given off at the negative plate and oxygen at the positive. These gasses result from the decomposition of water. A battery is gassing and using water, because it is being charged at a higher rate than it can accept.

When the battery is charged, however, it is a good idea to allow the battery to gas for some time. This allows the acid to mix and keeps it mixed evenly through the battery. In fact, in deep cycle applications, a forced overcharge, called an equalization charge, will balance the specific gravity of the acid throughout the battery cells.

Equalization of flooded cell lead-acid batteries

Large flooded-cell lead-acid batteries require an equalization charge every twenty or so discharge/charge cycles. Most people who live off grid equalize their battery banks once per month when the generator is running.

So, just what is an equalization charge? It's a controlled over charging of the battery. The battery is allowed to charge until the terminal voltage reaches 2.5 V per cell or 15 V for a nominal 12 V battery. Caution must be taken that any loads connected to the battery being equalized are not damaged by the higher than normal battery terminal voltage.

The equalization forces the battery's electrolyte specific gravity to stabilize from cell to cell. After you equalize your battery, a check with a hydrometer can pin point any problem cells in the battery.

You should only equalize flooded cell lead-acid batteries. NEVER equalize a sealed lead-acid battery. Some absorbed glass mat (AGM) batteries maybe equalized, but only if the manufacture okays the process. For the majority of us, do not equalize any sealed battery.

Specific gravity in lead-acid batteries

Because the specific gravity changes during discharge and recharge, you can get a fairly good idea of the state of charge of the battery by measuring the specific gravity. You check the specific gravity with a device called a hydrometer. You suck up some electrolyte and read the results on the bulb that floats in the electrolyte.

- A fully charged lead-acid battery at 80°F has a specific gravity of 1.265
- A fully discharged lead-acid battery at 80°F has a specific gravity of 1.120

In tropical climates, it's common to reduce the specific gravity to 1.225 for a fully charged battery. In heavy deep cycle use and in very cold climates, the specific gravity is usually increased to 1.300.

Hydrometer reading can be a bit misleading. The specific gravity does not increase or decrease on a linear scale. You can charge and charge and charge some more and take reading all the time and see nothing happening on the hydrometer. Then all at once, bang! The hydrometer pops up to 1.256.

Sealed and Absorbed Glass Mat Batteries

What I have been talking about is the flooded-cell lead-acid battery. As the name implies, the plates are submerged into the liquid electrolyte. In the sealed lead-acid battery or absorbed glass mat battery the electrolyte is suspended in a gel.

Sealed-lead batteries, or Absorbed Glass Mat (AGM), first arrived in the early 1970s. The main goal was to create a cleaner and safer version of the battery. With no liquid to spill or slush around inside, the gel cell was able to operate in any position, even upside down.

This helped make the battery (almost) maintenance free, and made it popular in a lot of different applications, like computer UPS, rechargeable hand tools, security alarms, and so on.

Inside a sealed lead-acid battery the electrodes are wound around each other in a spiral pattern with a fibrous glass mat as a separator. This means that the battery needs much less electrolyte, since it is either combined with the electrodes or absorbed by the separator. In addition it also allows gases to move around freely inside the cell, which means that they are given the chance to recombine with the opposite electrode. So even though the battery is "sealed" there are one-way vents that will allow the gasses produced by charging to vent. Because even sealed batteries must vent, never place them in an airtight box while charging or discharging.

The sealed lead-acid battery works the same way as the flooded cell battery. The difference is how the electrolyte is gelled and not liquid.

We need to discuss the sealed lead-acid battery or gel cell a bit closer. Although the chemistry is the same as a flooded-cell lead-acid battery, these guys have a life of their own.

I don't know of any ham radio operators that do not have at least a handfull of these batteries in the shack. The gel cell is a lead-acid battery with a gelled electrolyte maintenance free lead-dioxide battery. (The word gel/cell® was the trademark name owned by the Globe battery division of Johnson Controls. The trademark is no longer active, however, so we just call them gel cells.)

Although the chemistry is the same, gel cells can be

used in two different applications. Johnson Controls states there are two types: type "A" and type "B".

The type "A" is used in standby applications and is designed for four to six years of continuous charging. This type "A" use can be found in uninterruptible power systems, standby power systems and emergency lightings systems among others. This type of system keeps a constant charging voltage applied to the gel cell.

The type "B" gel cell is designed to provide three to five years of service with between 300 and 500 normal discharge and charge cycles. You see this kind in applications such as emergency communications equipment, portable lighting systems and standby power systems.

Understand that the chemistry remains the same between the two types; it the kind of application in which they are used that makes the difference.

Charging the gel cell is the same as with flooded-cell lead-acid batteries with two exceptions. First, limit the charging current. The formula is 3 to 4 times the 20 hour rate. A typical 7.5 Ah battery at a 20 hour rate will produce 375 mAh. (7.5 divided by 20 equals 0.375) So, 375 mAh times 3 equals 1.125 A initial charging current. At 4 times the charging rate the initial charging current will be 1.5 A.

NEVER equalize a gel cell or other sealed lead-acid battery. When charging the gel cell battery in type "B" applications, charge the battery to 2.4 V per cell or 14.4 V for a nominal 12 V battery. When the battery is fully charged, as indicated by the end of charge current dropping, the charger should be disconnected or set to a float voltage of 2.25 V to 2.3 V per cell or 13.5 V to 13.8 V for a nominal 12 V battery.

In type "A" applications, with the battery under a continuous charging voltage, the battery can be left connected for months on end provided the voltage per cell is maintained between 2.25 V and 2.30 V per cell or 13.5 V to 13.8 V.

High Current Discharges With The Lead-Acid Battery

When the flooded cell lead-acid battery is subject to a very high rate of discharge, there is a strange phenomenon that occurs.

The area surrounding the positive and negative plates quickly converts the electrolyte from sulfuric acid to water. This happens on an atomic level. This occurs because the electrolyte can not get into the pores of the plates.

I am sure you have noticed this exact effect. You're car's engine is flooded and you crank and crank the starter motor until the battery dies. You get out of the car, make a phone call to the AAA and then come back five minutes later. You try the starter one more time only to find the battery has "recharged itself." What has really happened is the acid was able to get into the plates and once again there's a chemical reaction to produce the required voltage.

That is why a car battery uses a lot of very thin plates. The more surface area the more exposure to the electrolyte and thus produce a much higher discharge current. They can produce enough power to crank over the largest engines. It's possible to discharge one of these batteries at thousands of amps for a several seconds. By the way, you need one cold cranking amp for every cubic inch of engine displacement.

Different Types of Lead-Acid Batteries

In the example above, you can see that a lot of paper-thin plates will produce a lot of current for a very short time. It's just the ticket for starting an automobile engine. In fact, this group of batteries is known as SLI or Starter, Lighting, Ignition.

Having that many paper-thin plates is a disadvantage too. If you were to draw smaller current for longer periods of time, those thin plates have a tendency to warp and just fall apart. In other words, they can't stand up to being discharged over time. They suffer from being discharged more than 20% of their capacity. They are not designed for deep cycle use.

Deep Cycle Batteries

The difference between a deep cycle battery and a starter battery is in the construction of the plates. Those paper thin plates the SLI battery needs to produce high current can't withstand the cycle use in applications that require power for a long time.

In the deep cycle battery, the plates are much thicker and thus able to hold up to the deep cycles. When a battery is cycled with long discharges, the plates tend to bend and flex. So, the thicker the plates, the deeper the discharge can be. The maximum depth of discharge should be limited to no more than 80% even with a battery designed for deep cycle applications.

So, what type of battery should you use in your emergency back up system? Well if you plan on running your

---

Only the Rich Can Afford to Buy Cheap Batteries

It's easy to get caught up in the sizing of your emergency battery system. In fact, it's really easy to find that the number of days of storage has increased into weeks of storage. But when you get done counting up the batteries, your check book keels over from a heart attack. Yup! Batteries are expensive. A large battery bank can often cost more than the solar panels to recharge them.

So, you start looking for bargains. You want cheap batteries. And if you look around you're sure to find something that appears to save you a bunch of change. Guess what? You get what you pay for. There's no such thing as "cheap batteries." Sure you can find blems, out of spec and odd capacity batteries on the market.

Lead-acid batteries are heavy, nasty and I don't get much enjoyment toting them around. If you put blems, out of spec, fallout or returned batteries into your system, you will end up replacing them in short order. So, unless you have lots of money, purchase a good brand name battery with a warranty. Because only the rich can afford to buy cheap batteries and then turn around and replace them every other year.

station for a long time then the best bet are the deep cycle batteries.

You can get by with an SLI battery **provided you understand that it will not last very long in a deep cycle application.** If you plan on running your station once or twice a year from a battery supply, then consider an SLI battery. On the other hand, if you plan on cycling the battery deeply and using it every weekend, then the only battery you should consider is the deep cycle battery.

As you can see, the difference between the two is the construction of the plates. Because the deep cycle battery has much thicker plates, it is not designed for high discharge currents. They normally won't supply the high current necessary to start an automobile engine.

Some deep cycle batteries have been altered a bit to allow them to perform as both SLI batteries and deep cycle batteries. These usually show up as marine batteries. They have enough high current discharge capacity to start the boat engine and with plates thick enough to provide a degree of deep cycle use.

When you pick up a deep cycle battery, you'll notice it's difference immediately. They're really heavy. All that extra lead adds up quickly. But don't be mislead. A heavy truck or "CAT" battery is not a deep cycle battery just because it is heavy. Just remember this rule of thumb: If you want to start something, use a car battery. If you want to run a load, use a deep cycle battery.

MORE BATTERY TECHNO BABBLE

When you start looking for a battery to operate your shack on battery power, you'll be hammered with all sorts of numbers and figures.

Among the most common figures handed to you are cold cranking amps and reserve capacity. Of course there's the amp hour rating of the battery.

Cold cranking amps is the number of amperes the battery can deliver at 0°F for 30 seconds and maintain a terminal voltage of 1.2 V per cell. That's 7.2 V under this load for 30 seconds at 0°F.

Reserve capacity is the amount of time a fully charged battery at 80°F can deliver 25 A until the terminal voltage reaches 1.75 V per cell. That equates to a battery terminal voltage of 10.5 V *at load*.

So, the higher the cold cranking amps, the more power the battery can deliver for a short period of time to the starter motor. A higher reserve capacity in minutes means the battery has a higher degree of deep cycle ability.

Amp Hours In Battery Storage

Batteries speak in amp hours. After you get past the cold cranking amps and reserve capacity, it all boils down to amp hours. In a nut shell, one amp hour means just that. One amp for one hour.

Batteries are normally rated at a 20 hour rate. So, a Trojan T-105 is rated at 220 Ah. This means that we can discharge this battery at 11 A for 20 hours before the terminal voltage reaches 10.5 V *under load*. I put those last two words in italics because it means that the battery is still supplying current, but the terminal voltage has dropped to 10.5 V.

The confusion comes when we try to draw more than 11 A from the battery. The higher the discharge current, the less amp hours you will get from the battery. Most of the larger battery manufacturers have plotted the discharge current versus time. If we were to discharge the T-105 at 20 A, we could only do that for say 14 hours. In fact, you can discharge the T-105 at 75 A for 90 minutes.

On the flip side of this coin, if you discharge the battery slower, you can increase the amp hour rating of the battery. If you discharge at 6 A, then you may have increased from 220 Ah to 300 Ah. These figures are for example only and each battery and each manufacturer has their own numbers.

Lead-Acid Terminal Voltages

We have become accustomed to seeing 13.8 V when we operate our low voltage gear from a gird-powered power supply. Batteries don't supply 13.8 V, but rather they are very happy sitting at 12.6 V. That is enough of a voltage drop from a power supply that you can see the difference on the wattmeter when running from a battery bank.

When a lead-acid battery is taken off of a charger, its terminal voltage is higher than 12.6 V. But even a slight load placed on the battery will remove this surface charge. The battery will then reveal it's true terminal voltage.

And while most of us are used to seeing a drop of only a few mV when we load our power supplies down from 1 A to 25 A, the lead-acid battery is another story.

In fact, if an engineer designed a power supply with the regulation of a lead-acid battery, he would have been fired. The higher the load current, the lower the terminal voltage will be. It can dip down quite a bit and yet the battery is still

This photo is enough to make any emergency power fellow drool. A pallet full of Trojan T-105 deep cycle golf cart batteries.

considered charged and in good shape. Remember that cold cranking test? The battery is still considered good when its terminal voltage is all the way down to 1.2 V per cell. Of course, it is producing 900 A when the test is being conducted. The battery is still good as long as a 12 V battery can maintain 7.2 V.

A lot of people use the battery terminal voltage to measure the amount of power in a battery. It's a good start, but the battery voltage can vary widely as loads are placed and removed from the battery.

When is a lead-acid battery considered dead? The answer is quite simple. The battery is considered depleted when it can no longer run the load. You have to be careful here as some radios will quit working while the battery has plenty of power left.

If you monitor the terminal voltage, a lead-acid battery is considered discharged when its terminal voltage is 10.5 V at rated load. In other words, if you are loading the battery at 11 A at a 20 hour rate, the battery is depleted when the terminal voltage hits 10.5 V

The same battery at rest would read 11.7 V. This is without any loads being placed on the battery. All the above assumes a battery temperature of 77°F.

This chart shows the relationship between battery terminal voltage and state of charge. The tests were all done at 77°F.

12.6 V or more	100%
12.4 V	75%
12.2 V	50%
12.0 V	25%
11.7 V or less	0%

Temperature Effects On A Lead-Acid Battery

Temperature has a great effect on the chemistry inside a lead-acid battery. Remember these are chemical machines. The colder they becomes the slower the machine runs. Likewise, if you heat the machine up, it will run faster.

When a lead-acid battery is very cold, you loose amp hour capacity. It's not lost, you just can't get to it. It's like placing a can of pop in the freezer and letting it freeze to just about a solid. You open the can and you can get some of the pop out, but the rest is still frozen. It's not that you lost what was in the can, you just can't get to it until the can warms. That same thing happens when the lead-acid battery becomes cold. You loose amp hour capacity but you can get it back when the battery warms up. You can also get more amp hours out of your battery if you increase its temperature. If you were to do this, you would shorten the life of the battery, however.

Temperature Compensation

While a fully charged lead-acid battery will not freeze, it's ability to take a charge is reduced. Likewise when the battery is hot, it will become charged quicker. To compensate for this some battery chargers and solar charge controllers will alter the final charging voltage. The standard is 3 mV per degree C. Since I have installed my battery bank in the garage, I have installed two temperature sensors; one

Batteries come charged and ready to go. Most will be a bit happier if you give them a slight charge before placing them in service.

for the charge controller and the other for the internal battery charger hiding inside the inverter.

Charging Lead-Acid Batteries

You can use either constant voltage or constant current charging schemes. In a constant voltage charge, the voltage is held steady and the battery is allowed to draw as much current as it requires. As the battery becomes charged, the current will drop to lower and lower values.

Constant current in its simplest form is like force feeding the battery. The battery is forced to receive a constant amount of current regardless of its need. It is also the most economical charger to build.

Trojan Battery recommends that a measured amp hour recharge is best. In its simplest form, you put more into the battery than you took out. Trojan recommends 12% more input than discharge current. You can also recharge less than or equal to C/10 times 1.12. C is the capacity of the battery at a 20 hour capacity of the battery.
- Limit current to C/10
- Recommend daily charge of 14.4 to 14.8 V
- Recommend equalizing voltage limit to 15.5 V. Hold for 2 hours then return to daily charge settings.

Charge float:
- Charge current upper limit of C/10
- Hold the upper voltage limit at 2.2 V per cell (13.2 V for a 12 V battery) at 80°F for a minimum of 12 hours to accomplish full charge. At less than 80°F, the voltage limit is 2.27 V per cell (13.62 V for a 12 V battery.) The battery can be left at the recommended float voltages for months.

I like to keep my battery bank charged to 14.5 V for

flooded cells and 14.2 V for SLA batteries. My nominal 48 V system then sits at 58 V when fully charged. When I equalize my battery bank, the terminal voltage reaches 62 V.

TESTING LEAD-ACID BATTERIES

While the hydrometer is a good tool for testing a battery along with a good digital voltmeter, there's nothing like a load tester. This device places a heavy load on the battery while it measures the terminal voltage. When it comes to testing batteries, the load tester is God! In fact, way back when, load testers were made to test just one cell at a time. The bad cell was located and it was removed from the battery and replaced with a new cell. This same technique is used in large batteries used in fork lifts.

Advantages and Disadvantages of Lead-Acid Batteries

While I went into detail on the lead-acid battery, it is far from perfect. But they do have a lot going from them.
- They're economical. They're cheap to make.
- They can be very tolerant to abuse.
- They can produce huge amounts of current for short periods of time.
- They can produce power for a long time.
- They are recycleable. Nearly the entire battery, from the case to the lead terminals can be recycled.
- They have a world-wide installation base.

They also have their share of problems. The biggest is the fact that they are made out of acid and lead — two rather nasty materials to say the least.

Another problem is the energy density. There are a lot of different chemistries that out-produce the lead-acid battery when it comes to energy density.

While all batteries can produce some gas, the lead-acid battery produces hydrogen. That can become explosive when in a confined area.

Here Harry is adding water to the batteries on the charge rack. Harry, where are your safety glasses?

Battery Maintenance

There are two areas you need to be concerned about. One is that lead-acid batteries require water to be added from time to time. The water should be distilled, deionized or other approved processed water every four to eight weeks. Never use tap water or mineral water. The impurities in the water will poison the battery.

The use of hydrocaps can reduce the amount of water required by the battery. Hydrocaps have a catalytic converter that recombines the hydrogen gas and oxygen back into water. They are very expensive and are used mainly in locations that make routine maintenance very expensive.

The tops of the battery should be cleaned with a weak solution of baking soda and water. Be sure you do not allow any of this solution to enter the battery via the caps. This mixture will neutralize the acid on the top of the battery. Gunk on the top of the battery will allow a current path to be produced between the two terminals. A clean battery will prevent such self discharge.

Every few weeks, you should check the hardware holding the cables onto the battery's terminals. They should be kept tight. A loose connection and high current can produce enough of a voltage drop that the heat produced will melt the terminals.

On batteries with a stud sticking out of the top of the terminals you'll find you can't keep them tight. This is known as lead creep. The stud

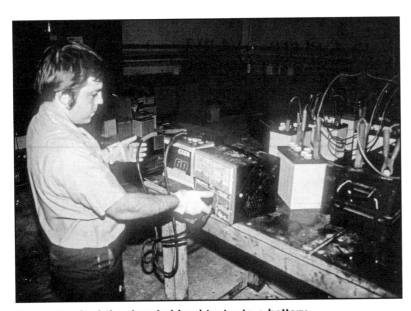

Harry attached the dreaded load tester to a battery.

Heavy cables and an air space between batteries is good practice

is embedded into the terminal at the time of manufacture. As you tighten down the hardware, you apply pressure on the stud. The lead will slowly "give" and after a few weeks the nut can become so loose you can turn it with your fingers. You tighten the nut again only to have the process repeat itself. After a while you can literally pull the stud out of the terminals. So, if it all possible, don't use the internal stud for your connections. The L-16 battery has terminals that allow a bolt to be passed through it. The use of clamp-on connections are good, too. Although more expensive from the get go, they avoid a lot of problems down the road.

I don't like to use grease or other goo on the terminals. While they do prevent green hairy things from growing on the battery terminals, gooey terminals attract dirt. I use a battery terminal paint. You can get a lifetime supply from any auto parts store.

Charging the Lead-Acid Battery, Part Two

Earlier I explained the internal workings of recharging a lead-acid battery. Now it's time to get our hands on the hardware.

There is an absolute amazing variety of battery chargers made today, from a simple 1 A trickle charger to 300 A (or more) chargers and boosters. For emergency power, all we need is something simple. A trip to your local auto parts store or discount store will do the trick.

If your needs are modest and you have the time for recharging, then a charger from 5 to 30 A is all you need. That's more than enough to recharge a set of T-105 batteries. Chargers with lower currents will just need a bit longer to run.

On the other end of the scale, there are chargers made for quickly replenishing the battery bank. You will see these in RV shops and marine stores. You'll pay much more for one of these chargers but they do have a lot of capacity. It's not at all uncommon to see charging currents from one of these listed at 50 A or more.

These chargers usually have high frequency switching type circuits. Do they produce RFI? Sure, but the higher-end units have fairly good shielding and filtering.

Avoid cheap, low-end chargers that use switching type circuits. They're cheap because they leave out a lot of the pieces parts required to prevent RFI. I picked up a small "computer controlled smart charger" from a discount store. When that guy is on, you cannot hear anything on any HF band. Yes, it does make that much RFI.

Sometimes the computer controlled chargers won't let you charge your battery bank. If they sense the battery bank is too low, too cold or some other parameter is not quite to their liking, they won't start the charging cycle.

Many dc to ac inverters also have built in chargers. The inverter that I have will charge my battery bank at upwards of 70 A. It sure makes the electric meter spin when that bad boy fires up. So, if you plan on recharging your batteries from an inverter or high capacity charger and run a generator, make sure the generator has enough capacity.

The power supply you have that is running your HF transceiver is not designed as a battery charger. It will work if you put one in service, but it won't be happy.

When a lead-acid battery is discharged, it can draw an amazing amount of current from the charger. This would place a power supply into current limiting for a long time, until the battery becomes charged and the current drops.

You should also be able to equalize your battery bank every so many cycles or at least once a month for large flooded cell batteries. Smaller chargers don't have the capacity to do this.

No matter what kind or type of charger you get, the best way to determine that the battery is charged is by watching the current. When discharged, the battery will draw high

Here is the charging station at Harris Battery.

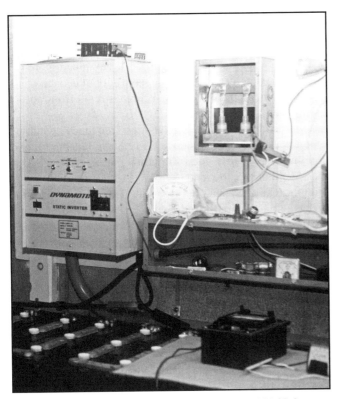

A 10 kW inverter being wired to this huge 110 V dc battery bank.

Batteries on Concrete

One of the cardinal sins to do at my dad's house was to place a car battery on a cement floor. I am telling you, demons would raise, the walls would bleed and you would be in deep dung if he caught you. For you see if you placed a lead-acid battery on a cement floor, the cement floor would "suck the power" right out of the battery. Or so says the old man.

In a way, he was right. To a point. A cement floor (especially in the basement) is usually cool and damp. This would cool the battery down to the point water would condense and lay on top of the battery. With a wet top, add in some dirt and some acid spray and you have a fairly nice electrical circuit set up between the two battery terminals. Over a month or so, the battery would self discharge very quickly and to the point it would be unusable without a recharge.

I sure don't want to battle demons so to this very day I never set a lead-acid battery on a cement floor. I use 4 × 4 wooden blocks to set the battery on.

current. As the battery becomes charged, the current will fall to a very low level, sometimes on the order of just a few amps on large batteries. It is at this time that the battery is about 90 to 95% charged. Allow the battery to continue to charge for a few more hours. This not only breaks up the sulfate crystals on the plates, but allows the electrolyte to gas. This gassing mixes the acid around the plates and prevents stratification of the electrolyte.

When the battery is gassing, it is producing hydrogen gas. It is at this stage of the charging cycle that is the most dangerous.

WHAT KILLS THE LEAD-ACID BATTERY?

One of the local battery dealers told me that over the many years he has been in the business, he has found batteries rarely die of old age: they're usually murdered!

I sure can't argue with that. As a matter of fact, I have killed my share, too. Usually, they die a slow lingering death.

The number one reason we murder lead-acid batteries is the lack of a complete recharge cycle. If you only charge the battery halfway, then the sulfate crystals that grow on the plates are not broken down. After a while these sulfate crystals become as hard as a rock and in effect kill a section of the plate. Once the battery has "sulfated" it's kaput! Completely charging the battery will prevent the sulfate crystals from choking the battery to death. A good equalization charge will help keep the battery from being sulfated.

Can you fix a battery that is sulfated? Well the answer is yes and no. According the people that manufacture lead-acid batteries, once the battery is in this stage, it's time for a new one.

There are people that market a thingie called a "desulphator." Now what's a desulphator? I'm glad you asked. A desulphator is an electronic device that sends a high voltage pulse that varies with frequency to the battery. The idea is that the sulfate crystals have a natural frequency and when they are hit (by the desulphator) with a voltage at that frequency, it will cause them to break up. This works much like the kidney stone destroyer that uses ultrasound to pound the kidney stones to dust.

Do they work? I am sure not going to tell you they do. I am also not going to say they don't. I have played with one on a really kaput battery and have not seen any marked improvement. That being said, I know of several hams that use them on all their battery powered equipment. Several of the desulphator manufacturers state their products are used by the "US military." I don't know if that means much to you, so I'll leave the desulphators and their use up to you.

Because of the high voltage switching going on with the desulphators, they make their share of RFI. There are none on any of my battery banks.

No matter what chemistry you're using, be it lead-acid, NiCd, NiMH or whatever, rechargeable does not mean forever. All rechargeable batteries have what is known as a cycle lifetime. Every time you discharge and then recharge the battery, you used up one cycle. Depending on the chemistry of the battery, your cycle life can be from 100 to over a 1000 cycles. Most lead-acid batteries can be cycled up to 500 times.

The kicker here is the depth of discharge. The deeper you discharge the battery, you'll end up with fewer discharge and recharge cycles. On the other hand shallow discharges will increase the number of possible cycles.

Those huge, heavy, lead-acid batteries used in float standby and only cycled from 10 to 15% can have a lifetime measured in decades.

For serious energy storage, these Douglas Battery deep cycle batteries await shipment.

INSTALLING A BACKUP BATTERY BANK

Although you could get by with other chemistries, such as NiCd, the number one battery chemistry to use is the old favorite, the lead-acid battery.

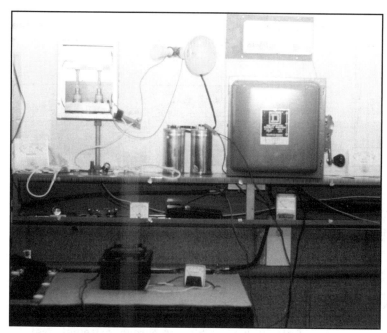

Ten 12 V lead-acid batteries being wired into a wind energy system.

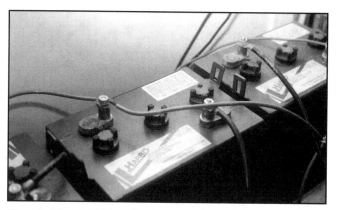

Here is how NOT to wire your batteries. The wire has no terminals and is just wrapped around the battery posts. The wire gauge is way too small as well.

While a small number of hams can get along quite nicely with a small battery tucked under the table, most emergency power back up systems rely on a more robust battery bank.

The larger the battery, the more amp hour capacity it can hold. There comes a point where you can no longer handle that big of a battery. A good example of this would be a fork-lift battery. They have current capacities of over 1500 Ah. They come in metal jacketed packages. And you need a fork-lift to move one.

So unless you have some way of man-handling a battery this large, we need to assemble our battery back up system using smaller capacity batteries. There are several things to keep in mind when assembling a battery bank.

1. Use all the same type of battery. Don't mixed NiCd and NiMH for example.

2. Use all the same capacity batteries. If you plan on using a group 27 deep cycle battery and you want to use four of them in parallel, then use four of the same type. Don't mix a 105 Ah battery with a 32 Ah battery in parallel.

3. The batteries should all be the same age. Don't mix newer batteries with older batteries. While good batteries are expensive and one would think you can add on as the budget allows, it's not a good idea.

4. *Never* use any type of diodes in your battery bank. Just parallel the batteries together and if necessary wire them in series to provide the required voltage. Look at rules 1 and 2 again. When you parallel batteries, they should always be the same type and have the same capacity.

When you wire you battery bank together, use heavy gauge wire. The smaller the number the larger the wire gauge. On systems that use inverters, 4 ought cable (wire size 0000) would be the minimum for connecting the batteries together and 2 ought cable (wire size 00) the minimum for running power to the inverter. Voltage drop, or I^2R losses, are a real problem when dealing with low voltage, high current circuits.

You need to include fuses as close to the bat-

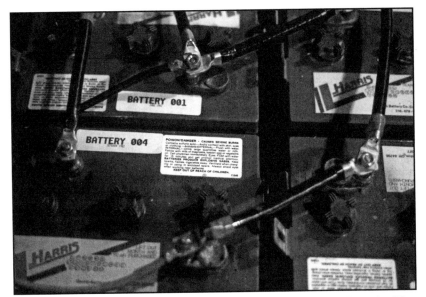

This is more like it. Heavy duty cable is terminated on each end with the correct ring terminal.

A set of eight Trojan L-16 batteries are shown in a battery coffin. Notice the insulation.

tery as possible. You not only need to protect your loads when things go wrong, but the wiring in between your loads and distribution busses.

When I install a large, high-amperage battery bank I always include a fuse as close to the battery as possible. This is a special T-class fuse rated for high current, low voltage interruption. In other words, it can open the circuit carrying high dc current with voltages up to 60 V. If this guy pops, you're going to say some things that just are not printable.

WHERE TO LOCATE YOUR BATTERY BANK

The best place is as close to the loads as possible. That's easy enough said, but harder to do than you think. Because of the voltage drop between the load and the battery, you need to run heavy wires and keep them as short as possible.

While some of us can get by with nothing more than a 105 Ah battery, larger capacity battery banks can take up a lot of room and weigh a ton.

If you don't require a large battery bank, you can keep a smaller capacity battery in a box under your operating table. You can get a plastic battery box made to hold one or two 105 Ah batteries at most RV or sporting goods shops. The box will keep cats, kids and critters out of harm's way. A pair of T-105 batteries would fit nicely in a wood enclosure.

If you're installing a large battery bank, then consider placing it out of doors in a battery coffin. Like the name implies, a battery coffin is nothing more than a large box to hold the batteries. Usually it is insulated to keep the batteries warm during the winter months. Be sure you put a lock on the battery box. You don't want your kids or the neighbor's kids in with the batteries.

Inside the battery coffin. The wires connecting the batteries is AWG 0000 welding cable. Notice the T-class fuse mounted above the batteries.

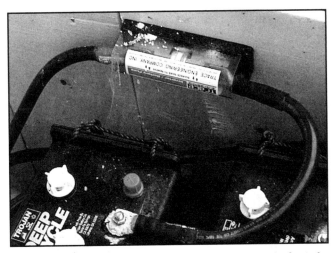

The T-class fuse is a high amperage, extremely fast dc rated fuse.

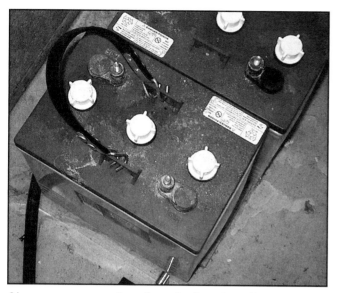

Always use a carrying strap designed to lift batteries.

Batteries in a String

When working with very large lead-acid battery banks, for some reason the end batteries in a string have a tendency to go bad before the others in the same string.

To prevent this, some installers have added extra cables that cross connect the batteries. In addition to the cables that place the batteries in series and parallel, an extra cable is run between the strings. In effect, you parallel the parallel strings.

If you don't want the extra expense of cross connecting the batteries, some experts suggest that you switch the end batteries with batteries in the middle of the string, much like rotating tires on your car. I don't know if this works or not. Since those L-16 batteries I use are kind of heavy, I'll leave them alone.

This line up of old worn out batteries is being removed.

Aside from a safety fuse inside the battery box, do not mount relays, switches or other devices that may produce sparks when activated.

Don't install a large number of lead-acid batteries in your basement. Nope, its not because of the hydrogen gas, it's according to *Mike's law of inverting weight*. It goes like this:

You carry eight T-105 batteries into your basement. With a little bit of loving care you can get about five to eight years of service out of them. When they have to be replaced, you are now eight years older, fifteen pounds heavier, and now you have to carry the old batteries *up* the stairs. Since you have to carry them up the stairs, the batteries now seem to weigh six times more than they did when you installed them. Been there, done that.

With the advent of high power MOSFETs, high efficient dc to ac inverters allow the batteries to be placed in a location that is close to the inverter. Then the output of the inverter is run to the radio equipment. I'll talk about inverters later in this book.

I wrote a simple spreadsheet to calculate the number of batteries I would require. The spreadsheet would allow me to input the cost of each battery and then it would show me the number of amp hours per dollar.

I found that although T-105 6-V, 220 Ah batteries would sometimes be cheaper on an amp hour base versus L-16, it would require more T-105 batteries.

More smaller capacity batteries means more cells to check every week and more cables (and thus more checking of those cables) to install. In the long run, it's best to use the largest capacity (in amp hours) battery instead of smaller batteries.

BATTERY SAFETY

No matter what kind of battery you are using, be it a small NiCd pack for your handheld or those eight monsters in my garage, they are dangerous. They both contain toxic material that if mistreated can hurt, blind or even kill you. Throw a NiCd battery pack in your pocket with some loose change and you'll see first hand.

Lead-acid batteries demand even more respect. They're heavy and contain acid. Even if you do nothing more than drop one, you're in deep trouble. One time I had two L-16 batteries tip over in my wife's minivan. The acid ate right through the carpet. I am still hearing about that!

You can hurt yourself just by trying to lift one of these brutes. The L-16s weigh in at about 100 pounds each.

SLAP YOUR BATTERY

What? Well any experienced solar panel installer will tell you to slap your battery terminals with the connector hooked up to the main feed wire. It works like this. That last connection to be made is usually the main feed going from the battery bank to the inverter or directly to the loads themselves. Normally you just grab the cable and bolt it onto the battery terminal. But what if you have left a screwdriver in the main disconnect box? That final connection could be your last. Sparks would fly, and more than likely the connector would weld to the terminal and then you would see the insulation on the wire begin to melt.

But if you take the connector and with a striking blow to the terminal slap it against the terminal, you'll still see sparks, but you'll avoid welding the connector to the terminal. It's an acquired skill but it's easy to do. Make slapping your battery a habit.

Keep these safety hints in mind at all times when working with lead-acid batteries:

• A lead-acid battery in particular can produce an almost unlimited amount of current if short circuited for a few seconds. A ring, a watchband or any other metal that comes between the two terminals will melt. Can you just imagine a ring with your finger in it touching the battery's terminals?

• At a minimum, always wear safety glasses. A full face shield would be even better.

• Use insulated tools so if you make contact with the other terminal, no sparks will fly.

• I always wear old clothes. Acid and cotton clothes makes for holes.

• Always vent the batteries to avoid hydrogen gas from building up.

• Needless to say, don't smoke, flick your Bic® or generate sparks while working on batteries.

Lead-acid batteries produce hydrogen gas when they are charging. It's the nature of the beast. There's nothing you can do to eliminate hydrogen from being produced. You can limit the amount of gas by keeping the charging current to a minimum and making sure you don't gas the battery excessively when the batteries become fully charged. Be mindfull that some gassing at the end of charge is a good thing. It keeps the acid mixed up and prevents stratification of the electrolyte.

Hydrogen gas production is a subject that should be addressed when installing your battery bank. The best place to put a large lead-acid battery bank is outside or in a garage. Some people have built special power sheds to hold batteries and inverters.

Small group 27 batteries and most AGM batteries will be happy to live under your operating desk. Just make sure you don't put them in a sealed (air tight) box. Give them some ventilation.

As a group, hams have taken it upon themselves to install deluxe anti-hydrogen battery containments. I've seen batteries housed in positive pressure boxes. I've seen battery installations that have plastic vent tubing installed in each cell to collect gas. I've seen all sorts of gadgets and thingamabobs to prevent hydrogen gas from collecting in a house.

Now don't misunderstand what I am about to say. If hydrogen gas builds up and there is a spark, you'll get the "Hinderberg" effect. For most of us, though, hydrogen gas build up is not an issue to be overly concerned about. Unless you have a super air tight house and four tons of batteries in the basement, hydrogen gas will filter out by itself.

On the same token, don't be making sparks at the battery's terminal when the battery is under charge either. A good dose of common sense around batteries goes a long way to a long and healthy life.

Batteries are the lifeblood of your emergency power backup system. When the lights go out, your batteries maybe the only source of power you have. Keeping your battery bank fully charged and well maintained will keep your station on the air.

Chapter 8

Systems for Emergency Power

For the last several chapters I have talked about generating electricity and storing it in batteries. We have looked at load sizing and battery storage. It's time to start planning a backup system. You will have to decide what level of emergency power you are looking for.

AUTOMATIC OR MANUAL

If you're really into an emergency backup system you may want a completely automated system. The power goes down, and you're on emergency backup in milliseconds. Automatic switch-over is slick. It can also become complicated very quickly. In addition, automation costs more money.

You can also switch over to battery and or generator power by unplugging and plugging into your backup system by hand. This manual system has a lot less parts that can go wrong. Each method has its own advantages and disadvantages.

Manual switch over is cheaper and works great. It's just a bit slower since you need to move connectors from the station power supply over to the battery bank or generator.

Let's look at the simplest setup you can have. I call this one the buddy system. You need just enough power to operate a 2 meter radio to talk to your buddy across town. Nothing more, nothing less. Just simple and easy to do.

THE BUDDY SYSTEM

The buddy system is by far the simplest and easiest way to get on the air when the power fails. There's nothing special about the setup, and it requires very little input from you, the user. It works like this: When the power goes out, for whatever reason, you can get on the air with a minimum amount of effort on your part. This system works best for getting a 2 meter radio on the air.

Here's What You Need

For the buddy system to work, you will need a battery and some extra radio power cables. The cables are radio specific so get the ones that will fit the radio you plan to use with the battery.

I suggest you install Anderson Powerpole® connectors

Here is how *not* to distribute low-voltage dc. I used RCA connectors. Needless to say, that was unsafe and caused too many problems. The meter monitored current being drawn. In the 1970s a lot of radios used RCA jacks for power input!

on all your radio equipment. You can make up an Anderson Powerpole®-to-radio patch cord to fit your radio. I have several old Heathkit 2 meter radios that have Anderson Powerpoles® on them. If nothing else, ARES and RACES have adopted these connectors. When you assemble your Powerpole connectors, orient them so the red connector is

Here is a complete QRP supply and automatic battery charger. It will provide both 13.8 V for a small load up to 1 A, and will also trickle charge a sealed lead acid battery. It is based on an LM317 adjustable regulator IC. Note the large heat sink on the LM317.

on the right and the "tongue" inside the connector body is to the top — the red right, tongue top orientation has become the standard way to assemble these connectors. That way you can interchange connections with other hams. This is also the orientation used on the various commercial power distribution panels.

In the buddy system, when the power goes out, you simply unplug the radio from its ac power supply and plug it into your battery. In most cases, a small 17 amp hour or 32 amp hour battery is all you need. When the power comes back on you connect your battery to a charger and refill the battery.

Let's make this system a bit more robust. Instead of the 32 Ah battery, go to your local discount store's automotive section. Get a cheap — and I mean cheap — car battery. You should be able to pick one up for under $25. While you're there pick up one of the small 1 A chargers, too.

You can place the battery in a wooden box, or better yet get one of those plastic battery holder boxes. Place this under your table.

Connect the trickle charger to the battery and then run a fused wire from the battery to a low-voltage power-distribution bus such as the West Mountain Radio RigRunner or MFJ DC Power Outlets. There are several other manufactures that sell similar low-voltage distribution strips.

The wire from the battery *must be fused* with an inline fuse *at* the battery. Do not depend on the fuse in the power bus. The wire needs to be protected between the battery and the power strip!

If you label this power strip for battery use only, all you need to do is unplug the loads from the main station power supply and plug them into the battery-supported power strip. It's simple and safe. The trickle charger will keep the battery charged and ready to go.

Ah, but you're thinking that's not a deep cycle battery. Yup! You're quite correct. The idea is to use this battery only during emergencies and only for several hours at a time. When the grid comes back on, the charger will once again refill the battery and hold it at full charge. After a dozen or so cycles, the battery will be toast. Get another one and recycle the old one. This is by far the cheapest, easiest way to put your station on emergency power. While not a deep discharge battery doing a deep discharge job, this set up works quite nicely.

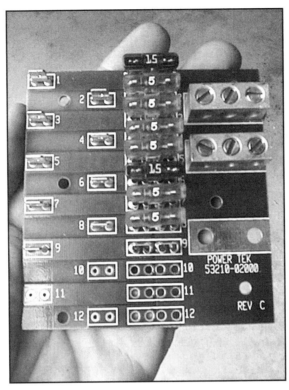

Here's a novel power strip. All Electronics Corporation was selling these for under $10. They lack the Anderson Powerpoles® but for installing in a shack or mobile home they are certainly inexpensive. Note that it is only for the positive connection. There is no ground bus on this board. Also, you will need to mount this board inside a plastic box and there are exposed copper traces on the other side as well. But what do you want for $10

An Improved System

Since most of our gear is set to run on 12 V, it's an easy matter to use your station power supply to keep a battery charged and ready to go. There are several ways of doing this.

The simplest is to parallel a 12 V lead-acid battery across the station power supply. When the power goes out, the battery takes over. When the power comes back on, the station power supply recharges the battery. All connections are made to a low-voltage distribution strip like one of the RigRunner panels.

There are a few problems with this. First, most station power supplies are not battery chargers. When the battery is discharged, it can demand all the current the supply can produce. This means the power supply will more than likely go into over-current protection, and it will also quickly overheat.

Switching-style power supplies can get upset quickly if you try to make them into battery chargers. Yes, there are battery chargers that use switching power supplies, but they began life as battery chargers, not power supplies.

Another bug in this setup is that most power supplies don't like to see voltage on their output terminals when they are turned off. You can modify some power supplies to prevent damage to them. The common Astron power supplies are easily modified. Contact Astron for instructions on how to do this with your Astron supply.

This is another example of a trickle charger and low current power supply. It is also based on an LM317 adjustable regulator.

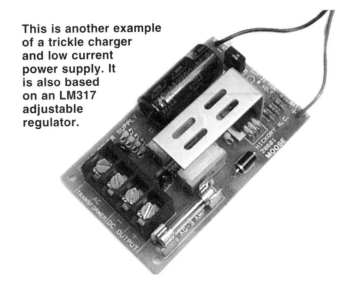

This is a another complete QRP supply and automatic battery charger.

Most of the power supplies today have some sort of over-voltage protection built in. Generally, the use of a "crowbar" over-voltage circuit consists of a Zener diode that monitors the output. If the output voltage exceeds 15 V or so, the zener diode conducts and fires an SCR. The SCR basically short circuits the output, popping the fuse in the process and thus protecting any attached loads from the 15+ volts. The only problem with this type of circuit is it can be tripped by a spike on the supply line, or noise on the input side of the transformer primary. These nuisance trips normally just mean you have to change the fuse every now and then.

With a battery connected in parallel with the power supply, if the over-voltage control trips, it will attempt to short the supply output — the same output that now has a battery connected to it. Now, when the crowbar circuit fires, not only will it short out the fuse, it will short out the battery, too. Since the battery has an almost unlimited amount of short-circuit current for several seconds, guess who's going to lose this contest? Now you can see why it is so important to fuse the battery as close to the terminals as possible.

I suggest if you go this route, you disable the over-voltage protection circuit in the power supply to avoid accidentally short-circuiting the battery. That opens us up for

a few more problems, however, as we shall see.

You can also install a large high-amperage diode in series between the battery and the power supply. The diode would prevent the battery from damaging the power supply when it is powered down. It will also prevent the over-voltage protection circuit from "seeing" the battery if the protection circuit trips. See **Figure 8.1**.

The diode will add some voltage drop, depending on whether you use a regular silicon diode or a Schottky diode. You can adjust most power supplies slightly to compensate for the voltage drop. If you do this, though, you must take into account that disconnecting the battery from the modified power supply will present a higher than normal voltage to your radio equipment.

When you increase the voltage from your power supply to overcome the voltage drop across the diode, you are really pushing the envelope of the over-voltage protection circuit. Most are set to fire at 15 V. If you figure on a standard 0.7 V drop across the diode, then your 13.8 V supply will need to be adjusted to 14.5 V. A Schottky diode will give you a slightly lower voltage point. The point being, at 14.5 V you're only a hair's breadth from the over-voltage crowbar firing. If you disconnect the over-voltage crowbar circuit, and the supply ever fails, the damage would be intensive. Now can you see why a power supply is not a very good battery charger?

Even the best diode will heat up at high current, so you must mount this diode on a large heat sink. If you plan to run high current, say 20 A or more, then select a diode with at least twice the current rating.

Replacing The Diode With A Relay

If you don't want to mess with the diode and heat sink, you can use a relay. But not just any relay will do. You want a relay called a *contactor*. This is a relay on steroids. It has contact ratings of hundreds of amperes. You need one that has a continuous coil rating. That means you can leave the coil on (and the contacts closed) for weeks on end if need be. You can get one of the contactor relays at just about any RV shop or boating supply store. Check with your local auto parts store, too.

Figure 8.1 — This is a very simple backup system. It uses the station power supply to float a deep cycle battery. Just remember that you should disconnect the over-voltage protection circuit in the power supply. You can leave out the diode if your power supply can handle having voltage on it's output terminals when it is off. Some can, some can't. The popular Astron power supplies can be modified to allow floating of a battery. Contact Astron for details concerning your unit.

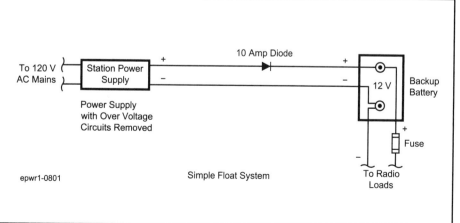

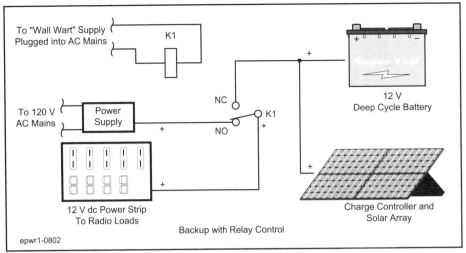

epwr1-0802

Backup with Relay Control

Figure 8.2 — This system uses a relay, K1, to direct power from the station power supply or the battery. In this example the battery can be charged by either a standalone 120 V ac charger or from a solar panel array. If you want to, you can have both the solar panel and ac charger connected to the battery at the same time. Normally, when there is power and the station supply is on line the output operats K1, pulling it in. When power fails, the relay drops out and power is drawn from the battery.

A mercury displacement relay would be an ideal choice here. Once again, you must have one with a continuous rating on the coil.

The contactor's contacts replace the diode. The best way to wire this is by using a small wall wart or other supply to run the contactor's coil. When the power grid is on, the contactor's contacts are closed and the station power supply is running the loads and charging the battery. When the power drops, the contacts will open, disconnecting the battery from the power supply. When power is back on, the contacts close once more and the battery will begin to recharge. See **Figure 8.2**.

By placing a rather large value electrolytic capacitor across the contactor's coil, the capacitor will keep the contacts closed during those times that the power flickers on and off. See **Figure 8.3**.

The above systems work and are simple to install and use. They do have some limits. The biggest one is the constant voltage placed on the battery for long periods of time. While not harmful to the battery, it can cause an excessive amount of water to be consumed. You reduce the amount of water used in the long term by adjusting the power supply so it keeps the battery sitting at about 12.8 to 13.5 V. The

exact voltage is not important, the idea is to keep the battery below 14 V.

Some of our equipment will get a bit fussy when it sees its supply voltage at 12.8 V. You must also take into account any wire losses between the battery and the loads.

There are still more ways to get on battery power. If you don't want to float the battery all the time from the station power supply, you can use a separate charger to keep the battery up to snuff, and leave the station power supply out of the equation. The way to do this is also twofold.

The simplest way is to use a pair of diodes. One diode prevents the battery from being charged by the station power supply. The other one prevents the power supply from seeing the battery.

What I've been talking about is called a *battery isolator*. They are normally seen as a device that isolates one battery from the other. You see these isolators in boats, campers and even in some automobiles. We can easily turn one around and use it as an isolator between the power supply and the loads. It's not at all unusual to see ratings of 30 to 50 A on most battery isolators.

While this configuration prevents the battery from being charged by the station power supply, you still have automatic switch over. You just have to provide the battery with its own charger.

This configuration works best with smaller loads. There's that pesky voltage drop across the diode when the loads are running from the battery. You can compensate by adjusting the power supply slightly higher when operating from commercial power. There's not much you can do to overcome the voltage drop across the diode when using power from the battery.

A QRP BACKUP SYSTEM

For under $10 you can build a QRP version of this circuit based around a common LM317 voltage regulator. The schematic is shown in **Figure 8.4**.

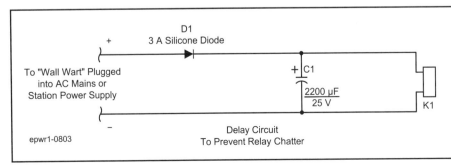

epwr1-0803

Delay Circuit
To Prevent Relay Chatter

Figure 8.3 — By adding a diode and a large-value capacitor across the relay coil, you can add a delay that will keep the relay from chattering when the power flickers on and off. There is nothing special about the values. Use what you have in the junk box. The higher the capacitance of C1, the longer the hold-in time.

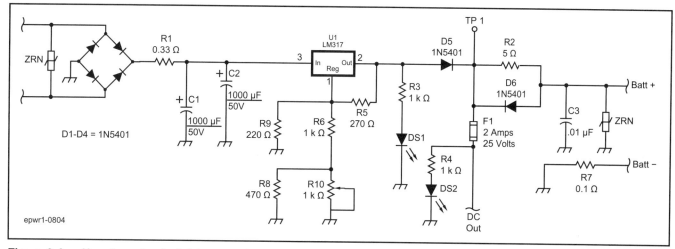

Figure 8.4 — Here is a simple 1 A battery charger and QRP power supply. Based on an LM317, this circuit will automatically switch from the 120 V ac mains to battery power. Diode D6 prevents the LM317 from seeing the battery when the power drops. DS1 is a power-on LED. DS2 lights when power is available from either battery or ac mains. Use any transformer from 14 to 18 V ac rated at least 2 A. Set up is simple. Adjust R10 while monitoring TP1 for 14 V. The voltage at DC Out should now be checked. It should be close to 13.8 V. If not, adjust R10 until it is. Connect a lead acid battery to the Battery terminals. The circuit works best with a battery between 4 and 24 amp hours.

It's very simple. The LM317 is adjusted to provide 13.8 V at the load terminals. The current limiting resistors limit the current into the battery. When the 120 V primary is removed, the load is instantly switched over to battery power by the diodes.

You could scale up the circuit to operate larger loads, but you have to remember that the LM317 is internally limited to about 1 A. The diodes are rated at 3 A. An LM350K would provide about 3 A of current to the loads and battery. You would need to increase the current rating of the transformer and bridge diodes, too.

If you try using a higher current regulator like the LM350K, you'll have to juggle the values of the current limiting resistor to prevent damage to the regulator. While the regulators are, for the most part, over-current protected, why let them cook for hours?

Adding a Contactor in Place of Switching Diodes

Once again our old standby, the relay, comes to our rescue. You can replace both or just one diode with a high-current contactor. You install the relay to connect the battery to your loads when the power goes out. The diode from the supply can remain if you want. Or, you can wire up a second relay to replace that diode. It can become a bit of a wiring mess with too many relays controlling too many things. That's one reason the use of steering diodes remain so popular.

CHARGING YOUR BATTERY

I explained battery chargers in the last chapter. For most of the systems described so far in this chapter, you can use a smaller battery charger than was indicated in Chapter 7.

Let's assume that you are not going to use your station power supply to float the battery bank. That's good. You have a set of back-to-back steering diodes to prevent the

battery and supply from seeing each other. That's good too.

So, to keep your backup battery charged, you have several methods to choose from.

1. Use a cheap car battery charger and just float it across the battery all the time.

2. Use the same cheap battery charger and let a timer control its operation.

In the first example, just get a cheap battery charger from the local discount store. Anything with a rating of between 1 and 6 A would be a good choice. You connect the charger to the battery and plug it in. The battery will charge when grid power is available and when the grid goes down, the battery will run the loads. Once grid power is restored, the charger will charge up the battery again. The small charging current means the battery won't be harmed, but it will take longer to recharge a completely discharged battery.

Although the standby current of a fully charged battery is quite low, if left in this condition, it's possible to cause the battery to use an excessive amount of water.

To prevent this, we need to look at the second example. Here, we change how the battery charger works with the battery. You can use a simple turn-on and then turn-off timer. Set the timer to run the charger for a few hours, such as between 3 AM and 7 AM every day. This will top off the battery from any self discharge and prevent excessive water usage by only charging a few hours each day instead of 24 hours a day.

You can even improve this method, too. Instead of a cheap one-period timer, replace it with a timer that allows multiple time periods per day and week. You could set this timer up to charge only an hour or so every day, and then run a longer time on the weekend.

The only down side to this is when you have a power outage. When power is restored, you should get the battery charged back up as quickly as possible. This not only prevents battery damage, but you may need to use the battery a few hours later if the power goes out again. During the August 2003 power outage, several communities had rolling blackouts while they tried to stabilize the electrical grid as it came back on.

EMERGENCY MICRO POWER SYSTEMS

This is a fancy way of saying "A battery in a box." The battery in a box is a portable example of the buddy system. When the grid goes down, and you need a little bit of power that is portable, then the emergency micro power system is perfect.

The one shown is my version of an emergency micro power system. It contains a 7.5 Ah battery and three ways of recharging the battery.

1. Using a 16 V ac wall wart

2. The internal 3909-based charger IC with built-in 120 V ac transformer. This is powered by external grid or generator power.

3. Solar power. Up to 5 A of solar panel current can be controlled by the built-in charge controller.

This unit is designed to be completely portable. Therefore the circuitry and control boards are placed into a Pelican case. This case is waterproof and the whole shebang will float.

Since the case is air tight when closed, there are no provisions for charging the battery when the case is closed. Even a gelled electrolytic battery produces gas when charging.

The loads are connected to my emergency micro power system using Anderson Powerpole® connectors. There are two outputs. Both are fused with resetting thermal chemical fuses. There is also an ATC 20 amp fuse wired directly to the internal battery.

Anderson Powerpole® connectors are used to connect the solar panel. In this case, instead of using a red connector for the positive lead from the solar panel, I chose to use yellow. This way, anyone using this emergency micro power system can plug the solar panel into the correct port.

This emergency micro power system has enough power to operate a small HF station at QRP power levels. It will also operate most VHF and UHF stations as well. In a pinch, it will provided an hour or so of 120 V ac using a small dc to ac inverter.

You don't have to get this fancy if you want to provide a source of portable power. A battery in a bag works just as well. The one shown in the photo is nothing more than the

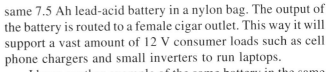

Here is my micro emergency power system. It's complete and waterproof. There are three ways to recharge the internal lead-acid battery.

same 7.5 Ah lead-acid battery in a nylon bag. The output of the battery is routed to a female cigar outlet. This way it will support a vast amount of 12 V consumer loads such as cell phone chargers and small inverters to run laptops.

I have another example of the same battery in the same nylon bag. This time, instead of the cigar outlet, I attached a pair of Anderson Powerpole® connectors. With this configuration, I can plug in to a low-voltage distribution strip and run a variety of small 12 V loads at once.

If nothing else, having a small lead-acid battery in a box or bag makes running portable HF a lot easier to do. Most of the smaller QRP HF radios will be quite happy running from a 7.5 Ah battery.

The battery in a box or emergency micro power system, will allow you to operate your hand held VHF/UHF FM transceivers for days or even weeks before requiring a recharge.

Never underestimate the importance of having a small, easy to carry source of low-voltage dc. You'll be amazed at how long you can operate from a small lead-acid battery.

PORTABLE EMERGENCY POWER ON A SLIGHTLY LARGER SCALE

Without digging out the generator, you can provide a fairly hefty amount of low-voltage power from your personal vehicle. Most stock alternators will provide up to 100 A for a short time.

In my Jeep, I have added a Mean Green high output alternator. At slightly over an idle, this

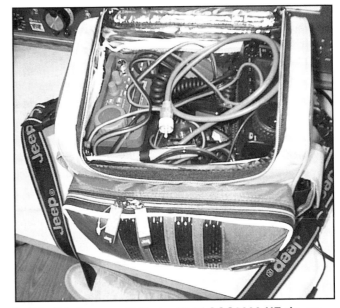

Inside this bag you'll find a 20 W SGC2020 HF rig, an ICOM 2-meter radio, an SGC 211 automatic antenna tuner and all the cables, connectors, spare parts and even two keys. Yup! Microphone's in there too. The side pockets also carry paper ARRL message pads and other odds and ends.

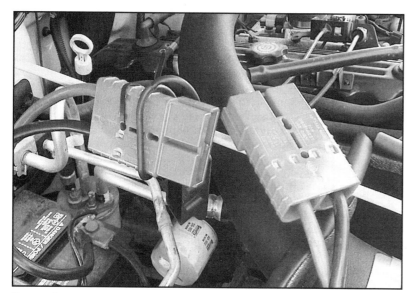

For serious connections to the high output alternator hiding in my Jeep, these 500A Anderson Powerpole® connectors do the job. I have another cable with the mating connector for charging an external battery from the Jeep alternator. That cable uses AWG 00 wire and is about 17 feet long. The set took a bite out of my check book, to the tune of about $300.

alternator will produce from 140 to 220 A. This is enough juice to run a high power dc to ac inverter up to about 1.5 kW.

Getting the power from the alternator is done by an industrial strength Anderson Powerpole®. This connector is rated at 500 A. I use one connector attached to the battery via a short run of AWG 0000 cable. I had a cable set made that will plug into the battery/alternator combination. I can use this to then run the power to whatever I need to operate. I also have a second cable set made that plugs into the battery. I use this pair as a standard "jumper cable" for jump starting a car with a kaput battery. This jumper cable set is supplied with a high quality alligator clamp that is rated at 600 A on each end. I can attach these alligator clamps to a second battery for a quick recharge.

For smaller loads I can run from my Jeep, I have a fused cable that is terminated with a set of Anderson Powerpoles®.

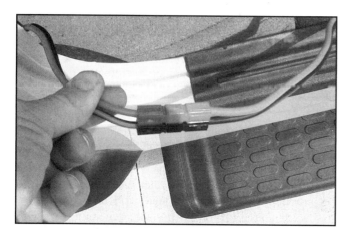

A set of 45 A rated Anderson Powerpole® connectors are wired directly to the battery in my Jeep. The line is fused at the battery with a 40 A MAXI fuse.

This cable can then be connected to a low-voltage distribution strip. There is enough power that I could run several HF stations, as well as numerous VHF and UHF stations. I could also use this output to charge up various handheld batteries.

I carry a small dc to ac inverter that is rated for 500 W to run those loads that require ac power. This system is great for running battery chargers for your laptops and cell phones.

I've taken small portable power one more step. I have a portable switching power supply installed in a Pelican case. This unit will provide upwards of 15 A to run most dc loads. While it is a bit short of running a 100 W HF rig, it will do just about 90% of full power. The best part of this is the ability to carry that much low-voltage dc in such a small lightweight package.

The unit comes with voltage and current meters built in. There are two dc outputs. Both are fused with resettable chemical fuses. An internal fan keeps the supply cool even when supplying maximum current. You can even press this system into charging lead-acid batteries if need be.

A PORTABLE SETUP ON STEROIDS

While most of us think portable emergency power systems are nothing more than a small battery and a radio, you can also assemble one that packs a punch!

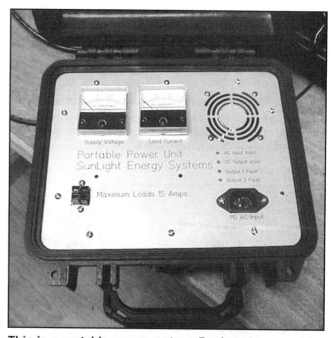

This is a portable power system. Designed to supply a nominal 13.8 V dc, it only weights five pounds. The output of the internal supply is monitored by both current and voltage analog meters. The system will produce about 15 A, just shy of operating a 100 W HF transceiver at full output.

Here is a low-voltage distribution panel and a small 12 V dc to 120 V ac inverter ready to be installed along the driver's seat in my Jeep. The inverter can be driven directly from the distribution bus, and can supply upwards of 300 W. That's enough for most small ac loads. In addition, with a 120 V ac power strip, I can operate a slew of wall warts for charging hand-held radio batteries during an emergency.

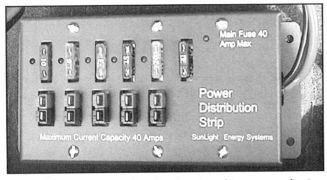

A low-voltage power distribution box is a convenient way to provide 12 V dc power connections for a variety of radios and accessories.

Since the inverter is so compact, and with the distribution strip next to it, I can operate and recharge dc loads and provide ac power quickly. They are mounted on the floor of my Jeep, next to the driver's seat.

Power Systems at WB8VGE

With all this talk about various methods of keeping one's station on the air, what do I use? Well as a matter of fact, I use a combination of several of the systems I described in this chapter.

If I just want to run a small HF radio on battery power during a short outage, I use the buddy system. I have several lead-acid deep cycle gell-cell batteries that I keep charged and ready to go at all times. These batteries range from a small 2.2 Ah to a larger 32 Ah unit. All are sealed so there's little to worry should the battery tip over on the table.

I can run a fairly large assortment of HF radios with a 7.5 Ah battery. If I want to operate longer or at higher power levels, I use a 17 Ah battery. I also have some 32 Ah gelled batteries on hand, too.

When the loads demand it, then the big one is used.

My emergency system consists of a Trace 5 kW dc to ac sine-wave inverter. The nominal operating voltage for this inverter is 48 V dc. From this inverter, generally about half of the house can be operated. I have the kitchen, living room and of course the radio room wired to the inverter. The deep well pump has its own 4.5 kW transformer to convert the 120 V ac from the inverter to the 240 V ac needed by the pump.

The inverter has its own transfer switch that will switch over from the grid to inverter in about 22 mS. In addition to the built-in transfer switch, a second transfer switch keeps the inverter from feeding back into the grid. This is not normally required, however, as the inverter has its own internal safety devices to prevent powering a dead grid circuit. The inverter must also pass the islanding test. A what? The inverter has enough capacity to power several homes if the grid were to go down and if the inverter were to be connected to the grid. Anti islanding software and internal circuits prevents this from happening. See the Inverter Grid Tie sidebar in chapter 9.

The battery bank consists of eight L-16HC lead-acid batteries wired in series. The nominal voltage of the battery bank is 48 V dc, but during charging it can reach as high as 58 V. The battery bank is installed in the garage, while the inverter and controls are located in the basement, in the radio room.

The batteries are charged with 2.5 kW worth of solar panels. I use a combination of Solarex MSX and Astro power modules. A 60 A charge controller monitors the batteries. The battery bank is temperature compensated by both the charge controller and the internal charger in the inverter.

DC power to the inverter is via a 175 A circuit breaker. The charge controller is protected by a 40A

You can for example toss a small 2.5 kW gasoline generator in the back of the truck and take off to where you're needed. Generators have gotten that small now. This is especially true with the inverter generators on the market.

But if you don't have a portable generator, then you should consider your own automobile. I outfitted my Jeep with a high-output alternator. This alternator will produce 200+ amps at slightly above an idle speed. Dual batteries

My second emergency power system. Notice how much more compact it has become. Also, notice the wind turbine controls are gone. Wind power did not work at my location.

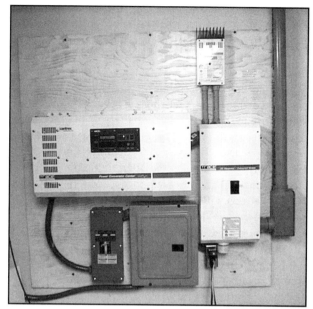

Here is my backup system as it appears in my shack. The main inverter is in the center and the 175 A disconnect is to the right. On top of the disconnect is a charge controller. Below and center is the manual transfer switch and sub panel that connects to the loads operated by the inverter.

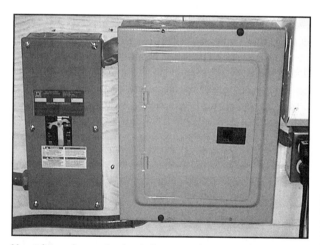

Here is a closer look at the transfer switch on the left and the sub panel conecting to the loads the inverter operates.

circuit breaker. All dc power circuits use UL approved wiring and use a common ground. All wiring is in approved boxes and all switch gear is rated for dc use. AC power is also controlled via circuit breakers and UL listed boxes and switch gear.

And how long will it run things? Provided there is enough sun, I could go forever. The system is sized to recharge the battery bank as quickly as possible, while having a combined storage of about four days. The loads and run times affect the storage capacity.

are fed from this alternator. The dual battery system allows me to have one battery run loads in the Jeep or externally via a high-current quick disconnect.

In fact, I had a set of cables made that will allow me to operate a high-current dc to ac inverter directly from the Jeep should the need ever arise. While I don't have an extra inverter laying around, I do have the option of operating it if need be.

I ran an extra pair of heavy duty supply cables from the battery to the driver's side of the Jeep. This way I can connect a low-voltage distribution strip directly to the vehicle. I could operate several 200 W HF stations this way.

While you don't need to go to this extreme, it's easy to do. Just fuse the power cables as close to the battery as you can. Run the wires though the firewall of the car and then terminate them into a set of Powerpole® connectors.

BATTERIES FOR YOUR BACKUP SYSTEM

As I mentioned in the buddy system at the beginning of this chapter, you can use a cheap "car" battery and get away with it. You can get by because you know up front that the battery will be used in a deep cycle application and you won't get more than a few dozen cycles before the battery is kaput.

When moving on to a more suitable battery backup system, we need to scrap that cheap car battery and go for an honest-to-goodness deep cycle battery.

The next step up is the group 27 battery. These "batteries on a rope" usually have a 105 Ah rating. One of these in a battery box placed under the operating table is a good way to get started. There's enough capacity to run most 100 W HF stations with ease. They're light weight enough so you can carry one downstairs. Or, if you want to, they are easily transported from one location to another. Hey, maybe that's why they come with a rope attached. Because they do have relatively short life of three to five years, using more than three or four of these batteries in a bank is hard on the check book.

The next step up is the one that I feel gives most hams the amount of power that will operate most of the gear for the longest period of time. I also feel this is the best setup for nearly 90% of all emergency power stations. It consists of two to four 6 V deep discharge golf cart batteries. Rated at 220 Ah at a 20-hour rate, these bad boys can deliver 11 A for 20 hours before needing to be recharged. Put in four, and you can draw 22 A for the same 20 hours. That's a lot of power in a small footprint.

There is a down side in using these batteries: they're heavy. Topping out at about 70 pounds each, they're something I would not want to lug around. Four of them would tip the scales at

My first emergency power back up system. On the right are four T-105 golf cart batteries. The batteries on the left are Edison cells.

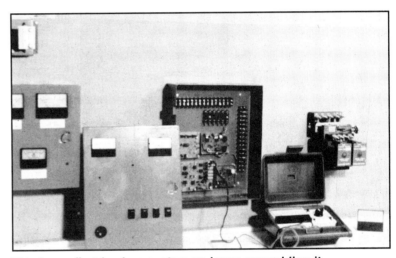

This is my first backup system as I was assembling it.

Here is a 10-pole low-voltage distribution strip I have been busy with.

This photo shows the set point meter and backup charger. When the solar and wind power were not enough to charge the batteries, the 120 V ac charger was turned on.

Power from the wind turbine and solar panels are routed to their own controllers.

LOW-VOLTAGE POWER DISTRIBUTION

With the popularity of the Anderson PowerPoles® for power connections, several companies have joined the bandwagon and have marketed distribution panels. These low-voltage distribution panels support currents up to 40 A. Because most ARES, RACES and even SKYWARN operators have adapted to these connectors, everything works together.

That's a long way from the Molex connectors I tried using. They worked, up to a point. The biggest drawback was the current limit of about 10 A. That was fine for most VHF and UHF equipment, but it's a not enough for a 200 W HF transceiver.

If you want to operate on emergency power, you must convert you equipment power cords to the Anderson PowerPoles®. That way, no matter what equipment you bring, you can "plug in" to whatever low voltage source of power there is. And if it's your turn to supply the power, then everyone can tap into your supply. It's a win-win situation for everyone.

Just make sure you have your connectors assembled to the ARES standard. And if you're going to go this route, make sure you have the proper contacts and the correct crimper tool.

A good place to start is by having a Powerpole party at your local ham club. Everyone brings in their own connectors and housings. Get the proper crimper tool from West Mountain Radio or from Anderson Power Products and crimp all night long. That way, everyone in the club can interchange radios and power sources with each other.

280 pounds. You will need a very heavy-duty battery enclosure to keep these guys safe.

They are kind of expensive too. Figure about $80 each for a well known brand name. Yes, you can get one cheaper, but remember that I said in Chapter 7, "Only the rich can afford cheap batteries."

I recommend the Trojan T-105 or the T-125 battery. Exide, New Castle, as well as other companies market golf cart batteries.

The next step up — and it's serious — is the use of L-16 batteries. They were designed to be used in floor scrubbers. The alternative energy people have made them the defacto standard in large-scale lead-acid battery storage.

The Trojan battery L-16 can provide 350 Ah at a 20-hour rate. The L-16HC will do 450 Ah at the same 20-hour rate.

These batteries are big, heavy and expensive. I have eight of them hiding in my garage. They provide storage for my solar electric backup system. The price tag for my battery bank is around $3000.

THE DOWN SIDE TO BATTERY BACKUP

While all the examples I have shown work and work quite nicely, they all suffer from one main problem: voltage drop. No matter how large of wire or size of the connection, there's that pesky I^2R loss inherent in any low-voltage system. The higher the current drawn from the battery, the higher the voltage drop.

In my first battery backup system, I used solar power to keep the battery bank charged. The battery bank consisted of 12 Trojan T-105 deep discharge 220 Ah flooded cell lead-acid batteries.

The batteries were fused at the battery box using two fuses. One fuse in both the positive and negative leads. The negative battery lead was not connected to the house ground. My system floated above ground at the battery box. Hindsight being what they say it is, that was not a good idea.

But anyway, from the fuse box and disconnect, AWG 00 wire was run to two large copper bus strips. They were mounted on a heavy piece of nylon about 0.75 inch thick. From these two

Although my system worked — and worked quite well — there are systems like this one. All homebrewed, it controls and maintains the batteries using solar and wind power.

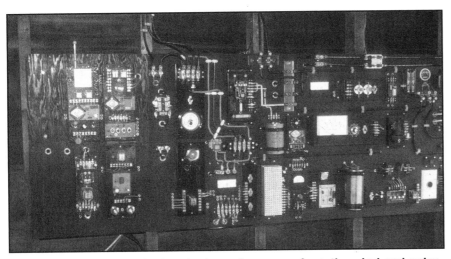

This panel is for conditioning the incoming power from the wind and solar panels. It also served as a backup system to the equipment shown in the previous photo.

copper bus bars, I connected all the loads. Since this was way before the popularity of the Anderson Powerpoles®, I used two pole Molex® connectors that were the standard for portable power systems at the time.

It was a mess! The Molex® connectors were only rated at 5 A. Large loads like my HF transceiver required much more curren than the connector could handle without melt-ing. So those heavy loads were con-nected directly to those copper bus bars. A fuse was included in each wire run to the load.

After all the high-current loads were connected to the bus bars, the smaller ones were routed to the bus bars via the Molex® connectors. Since these connectors were designed for hanging wires and not panel mounted, it was a cluster mess under the table.

After all the work I put into it, when I keyed down the 200 W input HF transceiver, I was seeing nearly a 0.75 V drop from the battery bank to the rear of the transceiver. While that's not much, it made a big difference on how much RF the solid state finals were sending to the antenna.

While you can do a little bit of cheating here and there to hide the symptoms, you can't really fix the problem of voltage drop with a snap of your fingers. You need to change directions and rethink how you plan on running your station from bat-teries, without the resulting voltage drop.

And folks that's called an inverter. So, in the next chap-ter I'll take a close look at converting our low-voltage dc into something we are familiar with: good old ac current.

Chapter 9

Inverters

One problem with running everything from a single 12 V dc power source is the power loss in the wires and connectors. This is known as I²R losses. Basically, electrons have a bit of difficulty moving through even the best conductor. That's why there is such a push to find a superconductor. In a superconductor, there is no resistance to electrical current. But we're a tad away from that happening.

So, in the meantime, to combat the losses in our wiring we continue to use ever increasing diameter wire. It's not at all unusual to see AWG 00 copper cables running from the batteries to the loads. The longer the wire runs, and the higher the current, the larger the wire must become. There comes a point that wire can no longer be used; copper bus bars are put to use instead. Running copper bus lines in your ham shack is neither practical nor affordable; the best thing to do is run a dc to ac inverter.

INVERTERS

It's not all that easy to make ac power from the juice of a battery. In fact, it's a science unto itself! The reason for all this fuss is the fact that direct current, as the name implies, is well, always on.

To increase voltage from one voltage source to another, a transformer is commonly used. Of course a transformer can increase or decrease a voltage, provided the power being supplied is ac. That's what makes a transformer work.

As you know, the ac in your home cycles back and forth 60 times a second. The waveform of the ac going to the transformer primary induces a voltage on the transformer secondary. Depending on how the transformer is wound, the output voltage will be more or less than the input voltage. (Okay, some transformers have an output voltage that is the same as the input voltage. They are called isolation transformers.) Usually transformers are wound with a simple ratio between the windings.

Direct current doesn't stop and change direction like ac does. So, there is no way to induce a voltage onto the secondary of the transformer, unless you turn the dc on and off.

Here's an example. To get the spark plugs to fire in the old family Buick, the points inside the distributor close and send current to the ignition coil. As the distributor's rotor turns, the points open. When the points open, the magnetic field in the primary winding of the ignition coil collapses and induces a voltage into the secondary winding. The secondary winding is designed so the voltage is increased to the point that a spark will jump across a sparkplug, exploding the air and gas inside the engine. There's more to it than just what I described, but the idea is basically the same. You need to turn the dc on and off to make the coil work.

Now, let's say we were to open and close that set of points really really fast. To do that, all we need to do is mechanically link our set of points to a moving arm like that in an ordinary doorbell buzzer. Do you know what you get when you're done? It's called a vibrator. Vibrator power supplies were used in the good old days to supply the needed 400 V dc for broadcast radios used in the early automobiles.

The vibrator would turn the primary of a trans-

This photo shows a 48 V backup system being installed.

former on and off at such a rate that the output would be a high voltage ac. The output would then be rectified and filtered before being used by the tubes in the radio.

While vibrator power supplies worked, to a point, they are full of problems. The number one fault came from the vibrator itself. Like tubes, vibrators were a plug-in affair. They wore out. Being mechanical, they would sometimes not start to vibrate! I remember my old man pounding on the dash of the Buick to get the vibrator to hum. And hum they did, for you see, that was another problem with vibrator supplies — they made a lot of noise. Not so much RFI, but you could really hear these guys singing. Vibrator supplies were none too efficient either. There was of course no voltage or frequency regulation. As the battery drained, you could hear the vibrator slow down.

Then along the way the transistor showed up and someone had the bright idea to replace the mechanical points with a pair of transistor switches. Not only would the switching speed be vastly improved, they were silent, too. And that is just what happened. As a matter of fact the circuit used became known as a multivibrator oscillator.

The transistorized high voltage inverters, as they were known, work fairly well, given the time period they came from. If you ran a Heathkit HW-101 or a Drake TR-3 mobile or from a battery supply, you were using a dc to dc inverter.

Along the way someone had the foresight to change how the transformer was wound. Instead of making 400 V dc for those 12AT7s why not make 120 V ac? The first dc to ac inverters worked, sorta, and after a fashion. They did in fact take the power from a 12 V battery and make usable 120 V ac power. The first units were very limited to the amount of power converted from the battery. Typically, you would see inverters with output capacities of between 200 and 500 W.

Let me divert just a little. These solid state inverters were not the first to take battery power and change it over to ac power. In fact, an inverter technology called a rotary inverter did the job first. The key to this process is the term "rotary," as the entire inverter was mechanical. A dc powered motor would spin an alternator, thus producing ac power in the process. They were also known as gensets, but once again, the marketing people had their say and rotary inverter sounded better than genset. Even though the rotary inverters were noisy and suffered from lack of efficiency, they did produce a sine wave output. They were not frequency controlled, nor was the output regulated. They made 120 V ac and that's just about all.

Old timers may remember the dynamotors. Like a rotary inverter, they had a dc motor that ran a dc generator. They produced the high voltage required for operating tube-based radios during World War II. While dynamotors produced dc, they did not produce usable ac power.

The output waveform from the solid state inverter was a square wave, which was just full of harmonic energy. The output frequency would run all over the place making the inverter useless if the load required a stable ac line frequency. You could not use one of the early dc to ac invert-

The battery bank for this installation was installed outdoors in an insulated box known as a battery coffin.

ers to operate audio taping equipment or other frequency sensitive loads.

They were also incredibly inefficient. At full load, they would be only 20 to 30% efficient in converting the battery power to ac power. If you needed a little bit of ac, though, and the nearest plug was miles away, they worked! About the only thing they were good at was running a light bulb or soldering iron. Some universal motors like the ones used in electric drills would operate from one of these inverters. (It is kind of whacky now, but the electric drill we use today would not run from these older inverters thanks to the variable speed function built in.) In a pinch, you might be able to run a blender or sewing machine. Early inverter technology was lacking. It was not until someone tossed the transistors in the dumpster and started to use power MOSFETS that inverters became much more useful. The results were simply amazing!

INVERTERS 101

Today we have a wide selection of dc to ac inverters available to use. They run from 50 W hand-held units to kilowatt units that mount on the wall. No matter what their size, they all work the same way: they take battery power and change it to alternating current.

Waveforms

There are three basic kinds of output waveforms produced by dc to ac inverters.

1. Square wave
2. Modified sine wave
3. Sine or pure wave

The square waveform is what you would find in the first generation inverters. As I mentioned earlier, the output was a square wave rich in harmonic content. Sensitive electronic equipment would not operate correctly. You won't find too many inverters that produce just a square wave output anymore.

The next, and most popular waveform is the modified sine wave. There really is no such thing as a modified sine wave. There's a lot of marketing and little engineering when you talk about modified sine wave inverters.

If you had a perfectly good sine wave to begin with, why would you modify it? That's like going to the doctor and saying, "I feel great. How about giving me something to make me feel lousy." But the marketing people have everyone convinced it's a modified sine wave, so that is what I will call it, too. As a matter of fact, once again marketing changed the name. Sometimes you will see a modified sine wave inverter called a "quasi sine wave."

No matter what you call it, the output waveform from a modified sine wave inverter has a few more steps in it. It's still a square wave, but those extra steps in the waveform will allow a lot more equipment to operate. If you have a modified sine wave inverter and suspect you need a true sine wave inverter, please note there is no filter or retrofit you can apply to the output of a modified sine wave inverter to clean up the output or turn it into a true sine wave inverter. What you have is what you've got.

There are some loads that won't work with a modified sine wave output. They may even go up in smoke. For those loads, you need a true, or pure, sine wave inverter.

TRUE SINE WAVE INVERTERS

As the name implies, the output is a sine wave, just like the stuff you get off of the grid. If you look at the output, however, you will still see steps in the output waveform. These steps are so close together they appear to the load as a sine wave.

As you can imagine, a true sine wave inverter can run just about any load you can mange to wire in. That includes deep well pumps, air compressors, heavy duty shop tools and even electric welders.

The problem with running heavy loads is the battery bank. Guess what? The larger the load the more battery bank you need. Now don't confuse this with run time. While run time is also dependent on battery bank size, the heavy loads require huge amounts of current to operate. Under full rated load the inverter can demand upwards of hundreds of amperes to start up the load. Suppose you want to start a five horsepower motor with an inverter. Yup! The inverter technology we have today will start that bad boy right up, *provided you have the necessary battery bank.*

That's so important, I put that last statement in italics. Running a heavy load is one thing, getting it started another. A popular sine wave inverter at full load requires 275 A at 24 V. Try to start a heavy load and the demand shoots upward to 700 A.

And yes, it is possible to run air conditioning units with an inverter. You need two things to do this. First, some of Bill Gates' money and second, several acres to hold your battery bank. Okay, it is possible, but just not practical.

THOSE SMALL HAND HELD INVERTERS

With technology making electronic devices smaller and smaller, there are now dc to ac inverters that provide upwards of 2000 W you can hold in your hand. These inverters do not use a heavy transformer. Therefore, they are upwards of ten times lighter yet provide the same power as heavier transformer-based inverters.

These inverters usually show up blister packaged and hanging on a peg at discount stores, home improvement stores and car parts outlets. Besides being lightweight, they're cheap, too.

These lightweight inverters bypass the conventional transformer by using switch-mode electronics. In effect these inverters take the battery power, oscillate it to a very high frequency and then reduce it down to 60 Hz and out to the load. As with most switching power supplies, there is an amazing amount of RFI produced in the process.

Depending on the price of the inverter, the noise produced may be so nasty as to make operation with a radio impossible. The cheaper the inverter, usually the more RFI it makes. To cheapen the inverter something has to go and that usually means the RFI filters. It might also mean a cheaper overall design was used! To sell a 700 W inverter for $39.95 something has to go. Clearly, this is a case of "You get what you pay for."

RUNNING LOADS ON AN INVERTER

The ac output waveform for many inverters is called a quasi-sine wave or a modified sine wave. It is a stepped waveform that is designed to have characteristics similar to the sine wave shape of utility power. A waveform of this type is suitable for most ac loads, including linear and switching power supplies used in electronic equipment, transformers and motors. The modified sine wave produced by the inverter is designed to have RMS (root mean square) voltage of 115 V, about the same as standard household power.

Most ac products run fine on modified sine wave inverters. True sine wave inverters are about two to three times as expensive per watt due to having more sophisticated design and manufacturing requirements, and more expensive components. As a result, most people prefer to use modified sine wave inverters if their applications allow it.

In general, any device that senses either voltage peaks or zero crossings could have problems when running from modified sine wave inverters. Devices such as these should be run from true sine wave inverters. Modified sine wave inverters tend to produce more RFI than a true sine wave inverter.

Electronics that modulate RF (radio frequency) signals on the ac line will not work and may be damaged. The X10 control modules won't work. You may notice hum or buzz in the audio of TVs, radios and satellite systems used with modified sine wave inverters. Audiophiles or professionals using sophisticated audio, remote measurement, surveillance or telemetry equipment should use a true sine wave inverter.

Examples of problem devices are motor speed controllers employing triacs, and some small battery rechargers that do not incorporate a transformer between the utility power and the load. To help you visualize this, if there is not a wall wart between the battery charger (or the battery in the device) and the ac plug, don't use a modified sine wave inverter.

Inside the battery coffin. Notice the thick foam insulation along the sides and top of the box. Not seen in the photo is the temperature compensation probe that the inverter's internal battery charger uses to adjust the charging

Two very common problem loads are electric shavers and emergency flashlights. Both of these items have batteries in them but connect directly into the wall to charge, without an external transformer. Don't use items like these with a modified sine wave inverter. If you do use a modified sine wave inverter with a transformer-less charger, your product will likely be damaged.

Garage door openers, laser printers and large strobes used in photography have all been reported as trouble loads for modified sine wave inverters; they either don't work at all or stop working entirely, so don't take a chance, use a true sine wave inverter instead.

As a general rule, products operating through an ac adapter will work fine from a modified sine wave inverter. These include laptops and cell phone chargers, video games, camcorder and digital camera chargers. Televisions generally work well; some VCRs with inexpensive power supplies run poorly. Consider switching to another brand of VCR in that case. A potential solution for RVers or off-grid cottagers is to purchase a small true sine wave inverter to run TV, VCR and audio equipment, and a larger modified sine wave inverter for the coffee maker, hair dryer and microwave.

I have also found that most ceiling fans have a tendency to buzz somewhat while running on a modified sine wave inverter. Those touch lights won't work or won't work correctly on a modified sine wave inverter.

As a general rule, you should never use an inverter of any kind to operate medical devices or life support equipment.

SIZING YOUR BATTERY BANK TO YOUR INVERTER

While I went to great lengths in Chapter 7 to describe sizing your battery bank, we need to once again look at battery sizing when dealing with an inverter. As a general

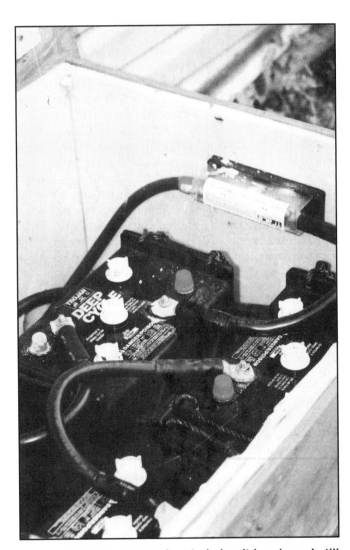

Even with all the proper dc-rated circuit breakers, I still like to add this class T fuse inside the box, as close to the battery as possible. Notice that the battery fuse is above the battery caps but not located at the very top of the box. That way, gasses and fumes won't damage the fuse or its holder.

Inverter Grid Tie

If you have been looking for a dc to ac inverter you no doubt heard the phrase "grid tie." That's the new buzz word in inverter technology. Here's how it works.

The inverter will normally operate from solar panels. These panels are usually wired for either 24, 48 or higher voltages. Some larger commercial systems use upwards of 600 V. Since the solar panels produce dc, a peak power tracker is normally used to match the output of the solar array to the inverter.

The inverter then looks at and monitors the incoming power from the grid. When the grid power is within specifications — frequency, power factor and voltage — the inverter starts producing power. When the inverter and grid are locked together, they are considered tied. The output of the inverter matches the grid power in all aspects.

When the two are tied together, any power being produced by the solar array is first sent to the loads being used by the household. If no one is home or the solar array is producing more power than needed, the excess is sent to the grid.

Utility companies really don't care for anyone connecting electrical generation equipment to their lines. (Do you remember how the phone companies yelled and screamed about hooking up a third party phone to their lines?) The biggest factor is safety. Linemen working on a supposedly dead line could be killed if the line were to be back fed by an inverter (or gas generator). I fully understand that position.

Another problem, besides the safety issue, was the quality of the power being produced by your system. Let's suppose that you hook up to the grid and because of the installation or the inverter itself, you end up feeding power back onto the grid when the power goes down. While you may not be feeding the whole city, your inverter has enough capacity to run the neighborhood. You have in effect "islanded" your neighborhood with power. If the power you are producing is not up to snuff and you end up cooking a rechargeable flashlight or burning down a house, the utility would be blamed even though the problem was caused by your equipment.

So there are very many guidelines and built-in safety devices and firmware built into a grid tie inverter. The anti-islanding standard was put into effect several years ago. All grid tie inverters now being sold must pass the strictest design rules before they are allowed to be marketed in this country.

Even with all those guidelines in place, when the technology became available to allow solar powered grid tie systems, the utility companies came up with all sorts of hurdles to jump over. There was miles of red tape to cut though.

Some went as far as requiring the customer to install two electric meters, and then charged the customer a monthly fee for both meters! Some utilities wanted ratcheting meters on the out-going power side. Others required at least a two million dollar insurance coverage.

Enter Avoided Cost

This is the price the electric company would pay you for any power generated by your solar electric system. In a nut shell, the avoided cost was the price of electricity minus the cost to transport it from there to here.

Whoa! That means that the super ZAP electric company will sell you power at 12 cents per kilowatt hour, but will only pay their avoided cost of 3 cents per kilowatt hour to you.

The government told the utilities they must allow you to hook up, but not how much they need to pay you for your power. All and all, it was a way for them to get you to say, "Ah the heck with it. It's not worth the trouble."

Net Billing

This is another buzz word you'll hear tossed around in the same sentence as grid tie. In its simplest form, net billing means that you have one electric meter. The power you consume at night spins the meter one way. During the day when you're tied to the grid, any surplus power spins the meter backwards. That's net billing.

Net billing is simple. You pay for what you use off of the grid and if you able to generate more than you need, you could end up with a check from the power company. Most just give you a credit towards the next billing cycle.

Most states have net billing. Some do not, others are working out the details. You will need to contact your local utility to find out where they stand.

Or You Can Go Gorilla!

That's the term for those that have connected to the grid in states that do not have net billing. In effect, they simply connect to the grid (using the latest technology that is safe) and spin the meter backwards.

Do I grid tie? Yup! I've pushed the required buttons on the inverter and did, in fact, watch the meter run backwards — for a while. I found that I used more power than my solar array can feed back to the grid. I decided I would just as soon keep the juice I am generating for a rainy day in the batteries.

Grid Tie Means No Power When the Grid Goes Down

There's a real problem with tying to the grid. Because of the safety features in today's modern inverters, when the grid goes down, so do you! Grid tie inverters do not use batteries for storage. No batteries, no power when the grid does go down. My SW5548 requires a small battery before it can interface with the grid, but the battery requirements are so small for grid tie, the battery would run the inverter for only a half hour or so.

The newer inverters that grid tie require *no* battery storage to operate. If the grid goes down, break out the candles!

rule of thumb, and assuming a nominal 12 V battery bank and inverter system, you can use this formula for calculating the overall battery current drain:

If you know the current drawn at 120 V ac, then the battery current at 12 V dc will be 10 times the ac current divided by the inverter efficiency.

The key here is that the ac load is measured in amps, not watts. So the label on the back of the TV may only show wattage and not current.

If the load is a 100 W bulb, then that's a little less than 1 A at 120 V ac. If this same light bulb is running from our inverter then the battery draw will be about 10 A dc. Throw in some conversion loss and the figure will approach 12 A dc.

Clearly, as you add up loads on your inverter, the battery bank must be capable of supplying the required power.

Because nearly every load we place on the inverter is measured in watts, and batteries speak in amp hours, we need to convert a few numbers. Let's look at a simple load sizing calculation using an inverter.

In any load sizing estimate, we need to find out how much power the load requires and for how long. In the case of ac loads, we will use amps instead of watts. One watt for one hour is one watt hour.

If our TV uses 100 W and we run it for four hours, we consumed 400 Wh. Likewise, burning a 40 W light bulb for six hours is 240 Wh.

Let's say for this example we end up using 5139 Wh. Now grab the calculator.

1. Multiply the total watt hours needed times the anticipated days of autonomy. That's the days of storage you want. Generally that will be between at least one and up to five days. To keep the battery bank within our budget, we will go with three days of storage. Three times 5139 is 15,417 Wh.

2. This next step sets the discharge of the battery. If you recall from the battery chapter, you should never discharge your lead-acid battery bank more than 50%. Keeping with that rule, we will set the discharge rate at 50%. To derive that number you take the storage watt hours divided by 50%. The result is 30,834 Wh.

3. We need to throw in some inverter efficiency loss and I'll set that at about 83%, which is a good figure with most inverters. The result is 37,000 Wh. This figure is the amount of power your battery bank needs in *watt hours*. But like I said, batteries speak in amp hours. To convert to amp hours you take the watt hours divided by the system voltage. In this example, I'll use 24 V. The result is 1542 Ah.

4. The number of batteries is then computed by taking the total amp hours and dividing that by the amp hours per battery. If we were to use 350 Ah Trojan L-16 batteries, we could get by with four batteries in parallel. Because the L-16 is rated at 6 V, we need four in series to get the required 24 V. Therefore our battery bank would consist of sixteen L-16 lead-acid batteries. Now at $350 a pop for an L-16 that's $5600. Ouch!

Inverters

The use of an inverter for grid tie is kind of romantic. Watching the electric meter spin backwards is novel. When the grid goes down, however, you had better break out the guitar and some candles because you'll be in the dark with everyone else.

Before the Y2K non-event of the millennium, the most popular setup was a small 1.5 to 2.5 kW modified sine wave inverter with a modest battery bank.

The idea was if rolling black outs hit your neighborhood, the battery bank would operate most of your critical loads for several hours. In a pinch, you could maybe run for a day on your battery bank. When the grid came back on, the internal battery charger would replenish the battery bank. What you really end up with was a whole house UPS. This setup was simple and relatively inexpensive.

A slew of these systems were sold and installed. The down side is some of the systems sold were done by people that had no idea what they were doing, other than making a lot of money off of the Y2K scare. So many of these installations just did not work correctly. In some areas of the country is may be possible to pick up one of these dc to ac inverter backup systems for next to nothing. It's worth looking into.

BATTERY CABLES AND CONNECTORS

There is one thing I hope you have seen so far and that's the amount of current a high power inverter can draw from the battery bank. The current is so high that the inductance of the battery cables can have a dramatic effect on how the inverter will run your loads.

We know that when current passes though a conductor, a magnetic field is set up around the conductor. As this magnetic field builds, it induces voltage in any conductor that is close by, and it induces a voltage in the original conductor. The voltage induced into the original conductor is called self inductance and tends to oppose the current that produced it.

The magnitude of the self induced voltage is proportional to the size of the loop formed by a wire. The larger the loop, the larger the self induced voltage. The positive and negative battery cables in a system are in reality only a single circuit and so the inductance of the battery circuit depends on how the cables are physically positioned or arranged with respect to one another.

If battery cables are separated by a distance, they have much more inductance that if there are close together. If the two battery cables were coaxial there would be virtually no induced current since the magnetic fields would cancel one another out. But you just can't find coax that will support the amount of current being drawn by the inverter, so it is back to ordinary wire.

We can cheat and approximate a coaxial cable by taping the cables together every four to six inches. When the cables are taped together, the magnetic fields around each battery cable tend to cancel each other. When the cables are separated the magnetic fields add together and increase the inductance of the battery cables.

This shows the amount of inductance in a pair of AWG 0000 battery cables running parallel to each other.

Distance between (inches)	Taped together	12	48
Inductance (µH)	3.3	6.0	8 - 9

Since the induced voltage in a conductor varies as the inductance times the rate of change of current in the inductor, the induced voltage may be three times greater that it would be if cables were not taped together. With all this inductance, it's quite possible that induced voltage spikes of thousands of volts may appear on the inverter input if the battery voltage were to be removed from the circuit. This might be the case if a circuit breaker opened under a heavy load, for example. High inductance on the battery cables can mean a loss of performance form the inverter, damage to the inverter and wasted battery power.

To avoid all these problems, tape your battery cables together every six inches or so. Mount them so they won't move and so they won't chafe against other electrical boxes. Use UL listed switch gear and fuses.

Use the proper size cables to begin with. At full power, my inverter can draw upwards of 175 A at 48 V. The minimum requirement is AWG 00 cable. For 2500 W inverters that run from a nominal 12 V system, 0000 cable is the minimum, and keep it under five feet total length out and back. That's two runs of wire (one positive and one negative) five feet long.

HIGH END SINE WAVE INVERTERS

The inverter I use in my system is a Trace 5548. Trace has changed its name to Xantrex, but they're still Trace to me.

This inverter will produce up to 5.5 kW and will surge to 10 kW for a short time. The eight L-16 batteries feeding power to the inverter do so with 0000 cable and a 200 A circuit breaker.

The SW5548 inverter I have will also do something really slick. It will back feed the grid. Yup! You can sell power back to the power company with this inverter. If your state has net billing, you just push a few buttons and spin the meter backwards. There is a price to be paid if you "grid tie" with an inverter. See the Inverter Grid Tie sidebar.

Because the inverter really produced power in sync with the grid, if you configure the inverter as a big UPS for the house, when the grid fails, the inverter goes on line within a half a cycle. The transfer relay requires 20 ms to switch.

The inverter also has a built-in battery charger that will produce upwards of 75 A to the battery bank. The battery charger and the inverter can track and compensate for temperature of the battery bank.

This inverter has the ability to automatically start a generator. It will also exercise the generator so it will always be ready to go. This inverter sells for about $4500. Xantrex has announced new lower prices, however.

What this inverter won't do is provide 240 V ac out of the box. To generate that voltage, requires you to place another inverter in series or use a step-up transformer. Since the step-up transformer is only $400 and a new inverter is ten times that amount, guess what one I am using. The step up transformer is used to operate my deep well pump. The transformer has a rating of 4.5 kW.

By adding two inverters, you would double your power. The added expense is a larger battery bank. You can also parallel two of the same inverters to again double your

Saving Money With a Grid Tie System

Every time the price of energy goes up, my phone starts ringing. The caller wants to tell the power companies to stuff it and has decided to go with solar power instead of the grid.

Well if you're trying to save money, you're looking at the wrong solution. It goes something like this. I'll buy some solar panels, a grid tie inverter and save a bunch of money on my electric bill. But guess what? You won't save money at all.

Let's look at the break down as simply as possible. If you are already connected to the power grid you pay for every kilowatt of power consumed. That rate will depend on where you live. If your power comes from hydro then your cost per kilowatt hour will be lower than my coal generated electricity. Currently I pay 5.4 cents per kilowatt hour.

A very basic grid tie system will consist of a solar array of say 1 kW and an inverter capable of connecting to the grid. A 1 kW solar array will consist of ten 100 W modules. At current prices of $600 per module that's six grand. A suitable inverter will set you back three grand. That's nine thousand dollars! And that price does not include installation, disconnects and other balance of system components. Nine thousand dollars buys a lot of power at five cents a kilowatt hour. For that matter, it buys a lot of electricity at 12 cents a kilowatt hour too!

Of course, that 1 kW solar array will not be enough to run much of your home. And if you have electrical loads such as an electric stove, electric water heater or electric clothes dryer, then you'll never see payback. Even without all those heavy electrical loads, getting a payback is really a stretch.

So, why do it in the first place? Well it's like the guy who likes to go Walleye fishing on Lake Erie. He buys a 28 foot boat with a large inboard engine. Then he adds a trolling motor, fish finders, GPS systems, radios, awnings and just about everything else you can stick in a boat.

Then he has to get a Ford F250 to pull the boat on a dual axle trailer. Gas on the lake is about $3.78 a gallon and then he has to pay all the license fees for the boat, trailer and truck. He does all this and more to fish for Walleye.

I can go drive up to the local store and get Walleye for $7 a pound. It's all ready to fry and enjoy. But the guy *really* likes to fish for Walleye.

It's the same for using a grid tie interface. While you'll more than likely never see payback, the knowledge that you're helping the planet in your back yard is the payback. Besides, it just feels good.

power. But you must also increase the battery bank to operate the larger inverters and loads. For you see, if you need more power than one inverter will handle, then you will of course need a battery bank that will handle 20 kW of peak load.

Because the SW5548 uses a transformer, the RFI it produces is minimal. Yes, there is some but for the most part it is hard to find on the HF radio. I have not noticed any on 6 meters and above. For all practical purposes, the SW5548 is RFI clean.

DRAWBACK OF THE HIGH END INVERTERS

Nothing is perfect and the SW series inverters are not exempt from problems. The biggest one that I can attest to is their bulk. The SW5548 that I have weights in at 117 pounds. It's big, it's heavy.

And they do break. It takes two people to remove the inverter from the wall. It's bolted down by a dozen large lag bolts. I can huff and puff and get the thing upstairs. But I can't reinstall it when I need to by myself.

At 117 pounds, the UPS shipping cost to the factory in Washington state is $145. One way. I know what you are thinking and no, you can't repair it yourself. Xantrex is really touchy about anyone opening up their products. Schematics? They may as well be on the Moon. So, when your inverter fails, and it will, it's off to the West Coast.

SMALL LIGHTWEIGHT INVERTERS

These lightweight powerhouses have come a long way. Yup, you can still get the cheap hamfest specials that produce more RFI than a spark transmitter, but the name brand inverters are simply amazing.

With no transformers inside, they are truly kW inverters that you can carry with you. Most are limited to 2000 W range because must operate on a nominal 12 V system. Power greater than 2 kW is hard to produce from a single 12 V battery.

A very simple inverter backup consists of a 350 W inverter and an ac power strip. This system is simple and easy to install in just about any ham shack.

These inverters are really happy out in the field or as a backup to a gas generator. You can run one of these inverters for a while before starting up the generator. They provide instant ac power from a battery.

You have to watch what you get when looking at this technology. They are all based on a modified square wave output design. They normally do not come equipped with internal battery chargers. And like I said, they can produce some outstanding RFI.

THE EFFICIENCY DRAGON

In the first part of this chapter I mentioned the I^2R losses when running directly from a battery to your 12 V loads. To eliminate these losses we can use an inverter that will convert the battery bank power into 120 V ac. Then we can use the station power supply running off of the inverter to operate the radios.

Now you're thinking that's plain crazy. Why invert 12 V to 120 V then back down to 12 V? Well that's the efficiency dragon. Dragons throughout time have lived with humans. We more or less get along. Yes, there is more inefficiency when we do all the conversion but we end up with better operating radio gear in the end.

As the battery bank drops under load, that new microprocessor controlled radio may cough and gag as the battery voltage sags. At some time there will be enough sag to cause SSB to start FMing and CW to take on a chirp. Some radios will just plain up and stop when the battery voltage falls below 12.5 V, still well within the battery's power reserve.

With the inverter running from the battery bank and the radios running off of their own power supply, everyone is happy. In a way, when you run your gas generator you end up doing the same thing. Instead of battery power, it's gasoline.

While all this conversion seems goofy, and yes it is inefficient, inverting from 12 V battery power to 120 V and then back to 12 V to run your radios is doable. So what is the bottom line? As my QRP buddies would say, "Sometimes even with QRP, the dragon wins."

PLAN FOR A BACKUP

It's easy to just purchase an inverter (or a gas generator for that matter) and then plan to run all your loads directly from the inverter. Today's modern electronic inverters are very complicated pieces of equipment. Like any other complex electronic gizmo, they break down. They usually break when we need them most.

I know first hand. My SW5548 went kaput for no known reason. Two friends later and $140 out of the check book, the inverter came back. It had a bad fuse on the FET control board. No, you can't replace it at home. It was a factory repair. So, for three weeks I was without my dc to ac inverter. Had the need arisen, I could still have gotten on the air. I can run the radio gear from a back up 12 V battery bank. I can't use the 48 V battery bank for the radios without the inverter, though. The only thing I could not operate without my inverter is the deep well pump. That's a problem I am currently working on. If you really want to keep on the air, then you need to think like NASA. You need backups for the back-ups.

A FEW OTHER THINGS

The input to your inverter is not normally protected against reverse voltage. If you hook up your inverter backwards, I can say with utmost certainty it will be toast.

High power inverters require large gauge wires from the battery bank to the inverter. Failure to use the correct gauge wire will more often than not cause a fire. Small gauge wire will simply melt when a heavy load is powered by the inverter. Even 750 W inverters should have the cigar lighter plug cut off and be wired directly to the battery (via a fuse and dc rated disconnect) for maximum efficiency. All wiring from the battery should be short, heavy and direct. While it is possible to operate a 2 kW inverter from your car battery, don't expect a full 2 kW worth of power out of the inverter.

I mentioned that some loads may or may not operate correctly when powered by the inverter. Try to operate everything you plan on using when the grid is down on your inverter now. Don't wait until you're in the dark to find out the inverter just fried your rechargeable flashlight.

There are some loads in your shack that require ac power. The antenna rotator is one that comes to mind. There are also computers and monitors, as well as printers that all need ac power. Make sure you try out each of these loads on inverter power before the need arises.

Unless your inverter is designed to operate in a grid tie configuration, *never* try to hook up the output of an inverter to the grid. Those cheap hand-held inverters will just go boom and then catch on fire. If you do manage to keep one from cooking, and without the proper safety designs built into the inverter, you can easily kill a lineman.

And speaking of cooking someone, that small dc to ac inverter you're holding in your hand can kill you! AC from the inverter is just as deadly as the stuff coming out of the wall. Do not under estimate the power that is being produced by even the smallest inverter.

SEARCHING FOR WATTS

Larger hard wired inverters have the ability to go into standby mode and just sit and watch for a load. They send a pulse of current out twice a second and if a load appears, they power up.

The inverters are looking for loads. What load the inverter sees can be set by the inverter. This is normally done by adjusting search wattage. It works most of the time. There are a few loads that can cause trouble for the search logic. A really good example is an ordinary light bulb. If you set the search wattage to 40 W, and you turn on a 40 W

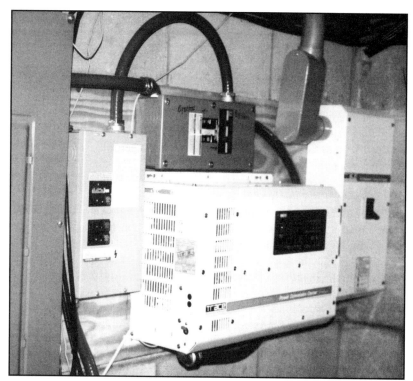

Here is another 48 V inverter being installed. This unit will be configured as a whole house uninterruptible power supply (UPS) system.

bulb the search will trip when it detects the light. As soon as the bulb turns on, however, the filament heats and the bulb may not draw enough power to keep the inverter on. The inverter shuts down only to find the now cool bulb and powers up again. The process repeats itself and as a result the light bulb flashes on and off as the inverter powers up and then shuts down. You can adjust the search wattage to either increase the wattage needed to operate the inverter or reduce the wattage so the inverter sees the smaller loads.

PHANTOM LOADS

Those wall wart power supplies will cause the search logic to go bonkers. If you set the search loads too high, then the wall warts won't be powered. Again, the fix is to either lower the search wattage until the loads are happy, or leave the inverter in the on position or simply turn everything off until the load is required.

A SIMPLE, CHEAP AND EASY INVERTER SETUP

While my system is hard wired into the electrical system of my house, you can have a simple inverter system that will power your shack without the wires, disconnects and whatnots. Here's how to do it. Decide what you want to run from the inverter. Your list may consists of the following:

1. HF power supply
2. Antenna rotator
3. High intensity light
4. Laptop computer

That's about 300 to 800 W of ac power. That's enough for a small high frequency inverter.

Now go to your local home improvement store and get a switched power strip. Get one that will have enough outlets so you can plug in all the stuff we have listed above. Then take some red or orange or other color of electrical tape and wrap some around the plug. This way, you can tell what cord belongs to what power strip behind your gear. If you're like me, there is more than a few lurking behind.

Just plug that strip into a 120 V outlet like you normally would. Life is good. When the grid goes down, here's what you do. First unplug the cord with the colored tape attached to it. Then plug it into the inverter. Supply power to the inverter and you're off and running. It's that simple! No fancy switching, no disconnects and no transfer switches. When the power comes back on, you remove power from the inverter, plug back into the grid and you're off and running.

Of course, you'll need a battery bank that can support an 800 W inverter and a battery charger that has enough oompf to get the battery bank charged up as soon as possible. Yes, you do have the inefficiency we mentioned earlier but you are on the air quickly and as simply as humanly possible.

You can take this setup one more step. Provided you can get power to the inverter, you can operate the inverter from your car battery. Remember, it is better to run the 120 V from the inverter to the shack than to run 12 V from the car into the shack.

As a side note, when dealing with high current dc, make sure you have the correct plugs and cables. Don't forget to install the necessary fuses between the battery bank and the inverter. Place your fuses as close to the battery as possible.

I have several small inverters. I use an Anderson Powerpole® connector that is rated for 120 A service. The smaller 300 W inverters are outfitted with 45 A Anderson Powerpoles®.

When you need ac power and you're miles from a power line, today's dc to ac inverters are just light years ahead of the stuff of only a decade ago. It's simply amazing what they can do. You can hold in one hand a 1 kW dc to ac inverter, and yet spend only $60 in the process.

Some units have internal battery chargers, others can start and stop an external generator. All provide peace of mind by knowing you can fire up a coffee pot when there's

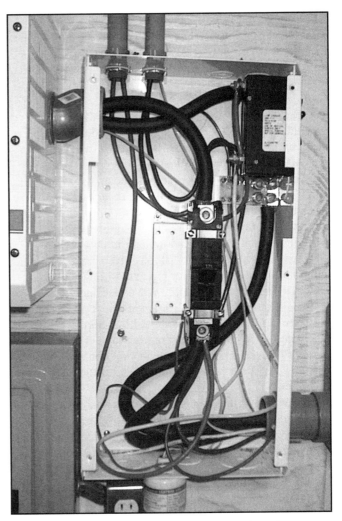

Inside the DC250 breaker box. There is a lot of space inside. That's because of the larger cables required for the batteries. The breaker mounted on the top and to the right is for the solar array. Just behind this breaker is the single point grounding block for the system.

no electrical grid power. With a small dc to ac inverter you can now do emergency repairs out in the field. From soldering a connector on a hunk of coax to running an oscilloscope are now possible with a inverter.

No matter how complex or simply your emergency back up power system is, you should plan on at least a few dc to ac inverters.

Chapter 10

Station Instrumentation

Well, after all the planning and the work, it's time to set up your system. You have decided by now how you want to control your power. It may be as simple as a car battery and some connectors or an automatic inverter-based switching system.

No matter what type of system you are going to use, it is important to have some way of monitoring what and how it is doing. Sure, you can use a portable DVM, but why? If your system is designed to operate your home station, then you should install some instrumentation to monitor what is going on.

BASIC INSTRUMENTATION

All you really need to monitor any of your emergency-backup-system parts is a voltmeter and an ammeter.

The Ammeter — Measuring Current

This meter shows current through a circuit. You can use it to monitor charging current *into* the battery from the solar panels or from an external battery charger. You can also use an ammeter to monitor current *out* of the battery as well. You'll find it very useful to monitor load current when troubleshooting your system.

If there is no current though your loads, then there has to be an open circuit between the load and the battery or power supply. Likewise, high current means you have left a load on, or you have a short circuit between your loads and the battery. The amount of current you want to monitor will have a bearing on the

Here is what happens when you try to push more current though an internal shunt than it was designed to carry. The internal shunt got so hot it melted through the plastic rear cap of the meter.

type of ammeter to use.

It's quite simple, really. The easiest way to measure current is to read the voltage drop across a known resistance. Using Ohm's law, you can then calculate the current through the circuit.

I am not going to get into measuring the internal resistance of moving needle meters. Instead I will use a very common meter value. The 50 mV meter means just what it says. Apply 50 mV across the meter terminals and the needle will deflect full scale.

To show one amp of current, we need to compute the value of the "shunt." A shunt resistor shunts or diverts 99.9% of the current to the load. Only a very small fraction of the current is sent to the meter.

One amp of current will require a shunt resistor of 0.05 Ω. When one amp of current is passed though this 0.05 Ω resistor, the result will be 50 mV. That's exactly what we need to send the meter's needle full scale. Remember, I am using the 50 mV analog meter as the reference.

Other meters may require different values of shunt resistors.

Using the same meter movement to show a full scale reading of ten amps the shunt resistor would need to be 0.005 Ω. Ten times the current, ten times less resistance.

Making small wattage shunt resistors is not that hard. Usually, all you need is a form of some sort — a large-wattage, high-resistance resistor will do — to wind some small gauge wire on. The amount of wire can be calculated given the amount of resistance per

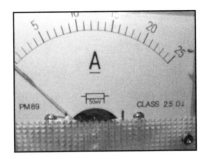

As is typical on most analog meters, while the meter's face shows 25 A, the meter's movement is really 50 mV.

A typical external shunt. This one is rated at 500 A at 50 mV. This means that when 500 A passes though this shunt, it will drop 50 mV.

foot for a given wire gauge. If you have the time, then make your own shunt resistors. I have found that 1 A to 10 A meters already have the required shunts built in. After you reach 10 A, then you need to use an external meter shunt.

External meter shunts are exactly what the name implies. The shunt resistor is external to the meter. Since the 50 mV meter is a standard used in commercial applications, external meter shunts that work with 50 mV meters are quite common. You can pick them up at surplus electronic shops like All Electronics. The shunts are standard too. Normally you will see shunts that work with 50 mV meters from 25 A to 500 A. These are called 50 mV shunts.

You can also find 100 mV shunts. Again, they come in the same ratings as the 50 mV shunts but are designed for use with 100 mV meters. 100 mV shunts also work with digital panel meters. Since most digital panel meters have an input of 200 mV, the 100 mV shunts work quite nicely. For example, 10 A current flowing though a 100 mV shunt would drop 0.100 V. On the digital panel meter, that would be displayed as 100. You need to move the decimal point on the digital panel meter to accurately display the result.

Armed with a few external shunts, you can monitor current through your system. One meter can show you current into your battery bank from the solar panels. Another will show you how many amps the loads are drawing.

External shunts also allow you to mount the shunt close to the loads and then run small gauge wire back to the meter itself. I use rotator-control cable. You can usually get at least four pair in a jacket. Use two or more pairs for current monitoring, via the external shunts, and one or more pairs for voltage monitoring.

While external shunts are great for keeping tabs on current in a dc circuit, they are not the first choice when monitoring current in an ac circuit. The easiest way to monitor ac current is to purchase an analog meter that will display the results directly. You can purchase an ac current meter from Mouser Electronics with ranges from 0-1 to 0-50 A ac. That's enough range to handle just about any load I can think of in a ham radio shack.

If you want to monitor larger ac currents, then you need a donut ac-current transformer. It looks just like a large toroid with windings in place. The wire carrying the load current is passed though the core. The result is an output of 5 A ac with 100 A ac current. A standard 0-5 A ac ammeter is then used to display the results.

The Voltmeter — Measuring Voltage

While a simple 0-15 V dc meter would be the first choice for most of us, there's something better out there. It's called an expanded voltmeter. A typical 0-15 V dc meter does not have the resolution to show you the most important narrow voltage band we are interested in. That voltage runs from 10 to 15 V dc. To display that range we need to expand the meter's range.

The schematic shown in **Figure 10.1** is what I have

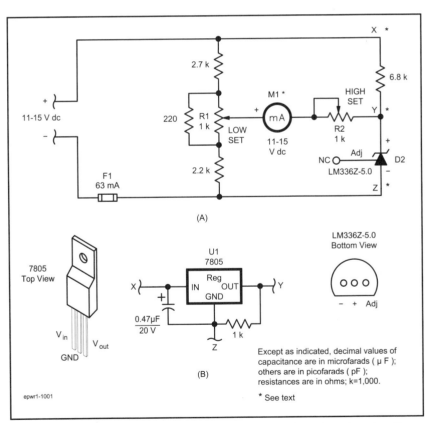

Figure 10.1 — Part A shows an expanded-range dc voltmeter circuit. Part B shows how you could use a 7805 voltage regulator IC instead of the LM336Z-5.0 reference Zener diode. The 7805 increases the current drain by about 8 mA, and is not reverse-polarity protected like the LM336Z is. The LM336Z is preferred.

Expanded Voltage Meter With Only Three Parts

Here's a slick circuit that only uses three parts. It only requires two IC regulators and 0-5 V meter. The hardest part to find is the 0-5 V meter. They are available from several sources, including All Electronics Corporation.

The circuit is completely linear and requires no calibration except to set the mechanical zeroing of the meter. By using a split-voltage reference system with floating output, the zero point of the voltage supplied to the meter equals the absolute sum of the two references. With the regulators shown in **Figure A**, the 0 to 5 V meter reads 10 to 15 V inputs.

For use with other voltages, select regulators for sums equaling the lowest voltage to be displayed. As an example, a 78L15 and a 79L05 will output a zero voltage at 20 V. A 5 V meter will then read from 20 to 25 V. You can make the circuit adjustable by substituting an LM317 adjustable regulator and the necessary resistors in place of the 78L05.

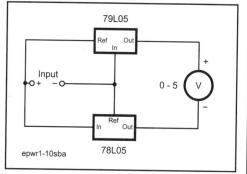

Figure A — This simple circuit shows a way to use two voltage regulator ICs and a 0-5 V meter to monitor the 10 to 15 V range as a battery voltage monitor.

been using for years.[1,2] Instead of a typical 15 V zener diode, it uses an LM336Z-5.0 precision voltage source. Toss in a few resistors and a pair of trimmers to set the low end and high end, and you're done. The values shown for the fixed resistors will need to be changed to match the meter you are planning to use. They may work for you, they may not. Use these values as a starting point.

To set the meter, you need to apply a known voltage of 10 V to the input. Set the low end trimmer so the meter just comes off the peg. Now set your power supply to 15 V and adjust the high end trimmer so the meter hits the top of the scale. You will more than likely have to do this adjustment several times as the two interact. Depending on the accuracy of the meter (most analog meters are 2% to 5% accurate) and your reference voltage source, the expanded voltmeter can be accurate to within 100 mV. You can modify the basic circuit to make an expanded ac voltmeter too. The schematic is shown in **Figure 10.2**.

I have two very old Heathkit expanded ac voltmeters. They have huge 6 inch meters you can read from across the room. If you ever have a chance to pick up one of these at a swap meet, buy it!

DIGITAL OR ANALOG METERS

So far I have mentioned only analog meters. Although it seems that everywhere we look we see a digital revolution, we still live in an analog world. I like analog meters.

Good analog meters are quite expensive. Mouser elec-

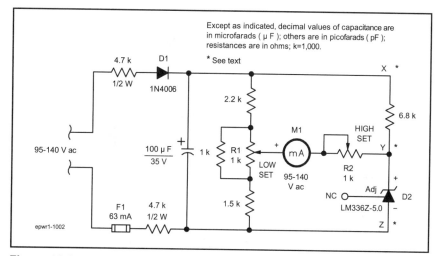

Figure 10.2 — This circuit shows how you can make an expanded-range ac voltmeter. You could also use the 7805 voltage regulator IC in this circuit, but it is not recommended.

tronics stocks Simpson analog meters. The average going price for a 2.5" meter is about $70. Good analog meters are expensive.

Check the surplus electronic suppliers for analog meters. All Electronics stocks a very nice selection. Made in China, they seem to be well constructed. They come as stand alone, or you can use external shunts to measure high dc current. Marlin Jones also offers a nice selection. And every now and then the Electronic Gold Mine will have some really impressive looking meters.

Most analog meters have the full scale deflection voltage printed down by the movement. Some will say 0-1 mA other will say 0-50 mV and so on. More than likely, some surplus meters will have odd and generally useless meter faces. There is shareware and free software that will allow you to scan in a meter face and then alter it. You then print out a new one and glue it to the meter.

Analog meters come in an almost never ending variety of shapes and sizes. The choice is up to you and the availability of the meters.

[1]J. Grebenkemper, "An Expanded-Range Voltmeter," *QST*, Dec 1992, pp 52-54. See also Feedback, *QST*, Feb 1993, p 78.

[2]J. Grebenkemper, "Expanded-Range DC and AC Voltmeters," Technical Correspondence, *QST*, May 1993, p 77. See also Feedback, *QST*, Aug 1993, p 69.

An expanded voltmeter is a great addition to your emergency power setup. This old Heathkit meter has been in my shack for years.

Analog meters come in all kinds of sizes and styles. All these meters display different values, but the basic movements are the same. In this case, they are all 100 mV movements.

Digital Meters

Digital panel meters *can* be quite accurate. You can find inexpensive meters, or you can pay big bucks. They can also be a pain in the neck to install in your system.

They come in several flavors. The most popular is the 200 mV, 3.5 digit LCD meter. I've seen these go for about $10 on the surplus scene. These cheap DPMs (digital panel meters) have several drawbacks. One, they are not overly accurate. In fact some of them are just plain inaccurate. They have a "set trimmer" that if you look at it, will change the accuracy of the digital panel meter.

If you get good first-run prime digital panel meters, be ready to drop some serious coin. It is not at all hard to find DPMs that run into $90 each. Of course, you get a lot more meter for your money when you compare a $60 meter to a $15 dollar meter. You get what you pay for.

Most DPMs require a separate power supply. This supply must be different from the supply you are trying to measure. This means you must install a supply to each meter you want in your system. While some DPMs will let you run more than one DPM from one separate supply, some won't.

That means if you want to measure battery voltage, battery load current, PV panel charging current and grid voltage, you will need a separate power supply for each digital panel meter. Most digital panel meters require +5 V dc. The LCD meters can run on a 9 V battery, too. But who wants to replace batteries all the time? Digital panel meters are also available with LED displays. These guys require about 200 mA of current at 5 V.

Working With Digital Panel Meters

Because the digital panel meters have a nominal input of 200 mV, you can't measure a higher voltage without a voltage divider. It's a simple circuit and only requires two resistors and one trimmer pot. While you can get by without the trimmer, you lose the ability to fine tune the voltage divider. I find it is easier to tune the digital panel meter to display the correct voltage by using an external trimmer rather than try to mess with the trimmer on the digital meter. Besides, the only way to really set the trimmer on the digital panel meter is to have an accurate 200 mV voltage source.

Remember, nearly all digital panel meters will require an isolated dc power supply. They can't display the voltage that is used to operate them.

To measure current, the digital panel meter works just about the same as an analog meter. The difference is in the input voltage. Instead of 50 mV, the digital panel meter has an input of 200 mV, so you need a shunt that will produce the required 200 mV. Lucky for us, these shunts are not hard to come by. A 50 A shunt can be purchased to match the needs of the DPM. So, a 50 A shunt that produces 50 mV at 50 A would display as 0.050 on the DPM. The leading zero is normally not blanked on cheap DPMs.

A 100 A current flowing through a 100 mV shunt would show .100 on the DPM. At 50 A, the display would show .050 once again. Notice I have left the left-most decimal point on. You can change what decimal point to turn on by shorting some pads on the DPM. With a quick touch of the soldering iron, the .050 becomes 050. and a 100 A current could be displayed as 100. just as easily.

LCD or LED Digital Panel Meters

You can't beat the low current requirements of the LCD display, and you can see the display in full sun. The LED display will wash out in sunlight, and they require a substantial amount of current to operate. I personally like the way the LED digital panel meters look. If you have several of them in your emergency power backup system, make sure you take into account the current required to operate them. Or, just put in a switch to turn them on and off as needed.

I have found that it is best to use one meter to display voltage and another one to display current. The input voltages from either a voltage divider or current shunt is so low that running it through a switch is not practical. There is not enough current to keep the contacts clean. So, a dirty switch will result in erroneous readings.

Two digital meters wired up to read incoming solar current and battery voltage. In this photo, the decimal points were not connected. The meters really display 14.0 V and 26.3 A of charging current.

WHAT TO MONITOR

I found that you should monitor at least two parameters in your system. One, of course, is battery voltage. The other is load current. If you are using solar panels to recharge your battery bank, then you should also monitor solar charging current. There's no need to monitor solar panel voltage, as it will be within a few tenths of a volt from the battery voltage. While it's nice to monitor ac load current, it's not that important in overall system operation. Monitoring grid voltage is a good idea, however.

CHARGE CONTROLLERS

Monitoring battery voltage and PV charging current is more often than not done by the charge controller. At the very least, it should have some way of telling you if the battery is charging and when the battery is full.

Some charge controllers can also measure the number of amp hours discharged from the battery bank and the number of amp hours being returned from the solar panels. The Trace C60 in my system will also keep track of the total watt hours generated by the solar panels as well as total amp hours.

INVERTERS

Those cheap hand-held inverters may not even have a green ON LED. The higher-end inverters usually have the ability to monitor several parameters such as battery voltage and ac load current.

Battery load current can only be displayed if an external shunt is used. Some inverters can't monitor battery load current even if the shunt is used. My Trace unit can't.

OTHER METERING SYSTEMS

There are several companies that market special monitoring systems. Besides the usual battery voltage and load current, they also give you a running total of available amp hours, percent of load, depth of discharge and a whole array of other parameters. They are a bit costly and they require several external shunts. Be careful when shopping for these units. Some come with the shunts while some do not. The price of the external shunts can add up quickly.

TOOLS TO TEST YOUR SYSTEM

It should be clear that when working on circuits that monitor voltage and current, the first line of defense would be a good digital volt-ohm meter. After you have spent all your hard earned money on solar panels and batteries, why buy a cheap meter? Get a good one. I prefer the Fluke brand. There are dozens of companies that make quality digital meters, however. Beckman, Fluke, Sperry and others all produce a meter that will stand up to abuse and are very accurate, too.

While you have a choice between an analog meter and a digital one, I prefer the digital voltmeter when working with dc systems. Sometimes you need to see that 0.3 V drop across a fuse holder. The digital meter will show you that voltage drop. An analog meter will, too; it's just hard to discern on the meter face.

For checking fuses and continuity, get a multimeter that has a continuity beeper. That way, you can trace wires for open circuits by just listening to the beep from the meter. This is a great time saver when looking for an open wire connection in a tight electrical box.

Along with your multimeter, a clamp-on current meter is worth its weight in solar panels. Most of the cheap clamp-on current meters will only read ac current. To read dc current requires a clamp-on meter designed especially for dc.

Most of the clamp-on current meters also do double duty as voltmeters. Some meters have a peak hold feature, which is great to have if you are measuring start-up current for an inverter.

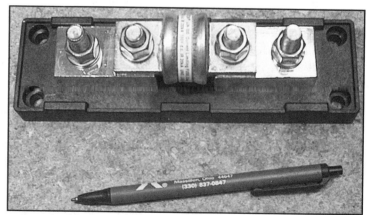

This is a class T fuse. It is designed for high current, low voltage systems. Note the large studs to connect the battery and loads.

Stuff That is Nice to Have, But You May Never Use

Most of the problems you face in your emergency backup system can be found and fixed with nothing more than your mulitmeter. But every now and then, when something is not quite right, an oscilloscope will come in handy. They're great when you need to look at the waveform of the ac output from an inverter. They can also be used to track down RFI that may be hiding on the dc power bus. You don't need a real fancy one. In fact a old and slow scope will work the best. Don't rush out and purchase one just for your system. They're nice to have, but you can certainly live without them.

BATTERY TOOLS

If you're into lead-acid batteries like I am, then you should have some battery-specific test gear. The first and foremost on your list should be a good hydrometer. Don't bother with one of the cheap discount store hydrometers. Get one made of glass, with a float inside. The ones made of plastic with the moving pointer are not at all accurate. Most glass hydrometers are temperature compensated, so the readings are accurate no matter what the electrolyte temperature.

Keep the hydrometer clean. You want to avoid contaminating the electrolyte with dirt, spiders, green goobers and other such things laying on top of your battery. That brings up a point. Why is all that stuff laying on top of your battery anyway? The idea is to prevent messing with the fine balance of chemistry going on inside the battery. I know one guy that uses distilled water to clean out his hydrometer before and after he tests each cell. It takes him a long time to do his battery bank.

It is a good idea to clean the hydrometer after you have used it, however. This prevents the acid from crystallizing on the float.

Be sure you write down the readings you take in your battery-service notebook. Hydrometer readings will help you spot trouble down the road with your battery bank, but only if you know the history of your battery bank to begin with.

Having said that, get yourself a spiral bound notebook and keep records. If nothing else, they can be a lot of fun to read over the years of your system's operation. It will also remind you just when you put that last set of batteries into service. It's amazing how we forget such things as time moves forward.

Once again, if battery power really impresses you, then you should pick up a load tester. It's nothing more than a really big resistor and a voltmeter. When you connect the tester across your battery, the resistor loads the battery down. The meter then displays the battery's terminal voltage. We know that when a lead-acid battery is loaded, its terminal voltage will drop. The battery must maintain at least one half of it's nominal terminal voltage under a specific load. In the case of a starter battery, the load may be as great as 800 A. The battery must maintain this current for, say, 15 seconds while keeping the its terminal voltage at

least 7 V. The example I just showed you will vary with the type of battery and the load impressed upon it, but the idea is the same. You load the battery down and then watch what happens to the terminal voltage.

Those dreaded load testers can be picked up from most auto supply stores or larger battery suppliers. I found some at Harbor Freight for about $40.

Way back when, you could get a cell tester. It looked like a BBQ fork with a heavy resistor between the prongs with a small meter across the resistor. The idea was to check each cell of the battery. Now that was back when the cell connections were made above the battery. I really don't know what you were supposed to do if you found a bad cell, other than drop some aspirin tablets into it. Then again, the bold could remove the bad cell and replace it with a new one. While that won't happen with today's smaller automobile batteries, the large batteries used in forklifts and trucks are checked for bad cells. Any that are found can be removed and repaired or replaced.

LOW VOLTAGE DISCONNECTS

After a while, you'll end up adding more and more loads to your battery bank. You'll be surprised how fast these loads add up. In a matter of a few months, you may have exceeded the capacity of your battery bank. What happens next is that you end up discharging your battery faster than you had planned. If you are like me and have a tendency to leave things running, you will wake up some morning and find the battery bank has completely discharged. The battery terminal voltage will be below the 10.5 V cutoff. Do this more than a handful of times, and you'll murder your battery bank. So what to do? You need what is called a *low voltage disconnect*, or LVD. Basically, it is nothing more than a electronic switch that constantly monitors the battery voltage. When the battery voltage reaches a predetermined point, the LVD disconnects any load connected to the LVD. The loads remain off until the battery has recharged to at least 50% of capacity. This prevents damage to the battery.

There are some drawbacks to a LVD. Most of the time, they are unrelenting. When the battery is discharged, they trip, and off the air you go. There's no resetting them while the battery is still low. High current LVDs use a heavy duty relay to control the loads. This relay is really a load and must be added into the over system loads you are planning on running. Granted, there are LVD that use power MOSFETS, but they are not without fault either. While a relay can withstand a short circuit on the load side of the LVD, a MOSFET junction is much faster than a fuse. MOSFET-based LVDs also use the MOSFET in a high-side switching arrangement. If you recall from Chapter 4, running a MOSFET as a high side switch requires that we raise the gate voltage above the switching rail. The use of a voltage charge pump can produce lots of RFI.

A good LVD will have several features.

1. Low standby current

2. A visual or audio (or both!) warnings of impending shut down

3. A delay before shutting down loads. This delay

maybe as short as a few seconds to several minutes. This delay will allow heavy loads such as an inverter to operate when staring up a motor or other load. A high power HF transmitter can pull the battery down during CW, so the delay will prevent the LVD from tripping during transmission.

4. A "one more time" reset. Some LVDs will allow you to reset them while the battery is still very low. It's like a snooze button on your clock radio. Pressing the reset will give you a few more minutes of run time before the LVD trips again.

MONITORING YOUR SYSTEM

I mentioned some of the parameters that you can monitor in your system earlier in this chapter. Don't go overboard. While it is impressive to guests in your shack, too many bells and whistles can complicate things. It can also make it look much more complex that it is. You may scare off another ham that has been

Where is the power switch? Or where is the manual? Don't go overboard when installing instrumentation to your system.

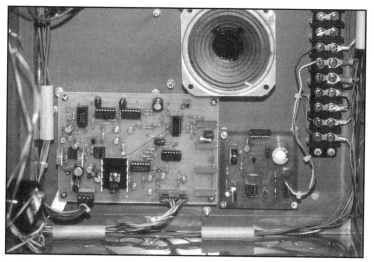

This project gave an audio warning of system errors in both the input from the solar panels and the output battery voltage. It was novel, but hardly worth the effort.

thinking of installing his own backup system.

So, keep it simple. I know, I've redone my system several times. At one stop along the way, I took a few steps back and what I saw was way over the top and gosh-awful complex. I had too many meters and alarms.

I went from monitoring every bit of information generated from the solar panels and wind mill — there were a dozen meters — down to just one. That one is mounted on the charge controller. Granted, the inverter has its own separate set of meters that monitor battery voltage, load current and output frequency as well as a slew of other things. But I use them mostly for troubleshooting my system.

I only monitor what is going in and what is going out. I know the current from the solar panels into the batteries, and from the batteries into the loads. The inverter can tell me the ac loads it is powering.

My current emergency backup system has a clean and uncluttered design. Although it takes up most of a 4 by 6 foot hunk of wall space, visitors don't notice it unless I point it out.

Chapter 11

Safety

This is perhaps the most important chapter in this entire book. Safety has the highest priority when dealing with energy. That goes for any kind of energy for that matter. It makes no difference if you're talking compressed air or a mechanical part under spring tension. When energy is released suddenly in an uncontrolled manner, things and people tend to get hurt. While I've been spreading the word on safety throughout this book, one more look won't hurt anyone.

I could save a tree or two and a lot of ink by condensing this entire chapter into one sentence.

"Don't do stupid stuff."

There. That's more than enough information on safety. You know there must be a DNA sequence deep down inside the human gene pool that keeps making us humans do stupid things. That's especially true around electricity.

I sure am not immune either. I have done my share of stupid things when working on electricity. I still have the burn mark on my finger when I wanted to see if the bleeder resistors on a power supply were running hot.

They were. They were also carrying 400 V dc. That got my attention big time! No more of that for me.

My favorite stupid thing I did was way back when I was just a small stinker. I wanted to see what electricity felt like. So, I unscrewed a light bulb and stuck my finger in the socket and turned the switch. Yup! I found out what electricity felt like that day.

Safety never stops. It's ongoing. We learn from experience. I've never stuck my finger in a light socket again.

I have found that when someone gets hurt, it's usually caused by four conditions.

1. Stupid stuff you did
2. Not thinking
3. You were in a rush
4. Something broke

Safety never stops. It's ongoing.

We already talked about stupid stuff. The not thinking part nearly cost me my life. Let me explain.

I was working on a home brew 813 HF amplifier. This amp was in two sections, the RF and high voltage decks. I was working the afternoon shift at the mill and had about 20 minutes before I had to get to work.

The high voltage to the 813 was coming from a collection of surplus transformers and 866 mercury vapor rectifiers. Unloaded, the power supply did about 2500 V dc. The amplifier was not working correctly and I was taking voltage readings, trying to find out what was up with the amplifier.

The high voltage was routed from the power supply deck to the RF deck by a hunk of RG-8U cable. The high voltage wire was attached to the RF deck with a push-in connector. You push the top of the connector, slide the wire in and then release the button. The dummy load I was using also had a hunk of RG-8U cable. It was hard to tell which cable was attached to the dummy load and which one was attached to the high voltage supply.

Since I was in a hurry, I powered down the power supply deck. I forgot to use my shorting stick to discharge the power supply capacitors. I disconnected the dummy load cable and then without thinking, pushed the button to release the high voltage supply wire.

Unknowingly, I had set myself up for the worst possible scenario. In that split second, I allowed 2000 V dc to flow from my left hand through my chest out to my right hand that was touching the RF deck. In a flash, I discharged the capacitor bank though my body. Had I left the power supply on, someone else's name would be on this book cover.

I never had a jolt like that pass though me, before or since. I was thrown to the floor and my chest tightened to the point I was not able to breathe for a few moments. No one was home with me. I was by myself. I was lucky.

I violated several common sense safety rules.

1. Never work alone when you're working on high voltage.

2. Always, always discharge all capacitors.

3. Slow down.

4. There should be nothing on your mind but the work at hand. No distractions.

5. Think about what you are going to do.

When I got home that evening I went downstairs and took the amplifier apart. I sold the parts and that was the last time I have ever worked on a homebrew amplifier. In fact, it is rare that I work on any high voltage HF amplifiers.

There are times when you do everything according to good basic safety guidelines, and you still get hurt. We call these accidents. Something broke. A wheel bearing breaks and the car runs off of the road. That's an accident that was not caused by something you did or did not do. Sometimes these accidents can be traced back to something stupid someone did.

By now you're thinking, "Okay Mike, what does all of this have to do with my emergency backup system?" Well, quite a bit actually. While it is true the shock hazard is much reduced when dealing with low voltage dc, the amount of current is the killer. Well okay, perhaps not a killer but with current measured in hundred of amperes, a dropped wrench across a couple of bare terminals can easily melt the wrench. The sudden flash of light generated and the molten metal hitting your hands and face can give you quite a scare. There's a possibility your reflex actions could put you in harm's way. That sudden jerk could send your hands into a live circuit.

That in itself is enough to give one a heart attack. The molten metal that will be sprayed can easily damage your eyes. In addition, when you jump up the bottom of the table holding your gear will give you nasty bump on the head. Yes, 12 V low-voltage systems can hurt you!

KEEP THE AC FROM THE DC

When wiring your emergency backup system, keep the ac circuits from the dc circuits. You don't only want to keep them apart, but keep them in their own separate electrical boxes as well. You don't want ac circuit breakers in the same fuse box as your dc breakers.

USE THE RIGHT FUSES

If you're dealing with high-current, low-voltage dc circuits, then you must use fuses and circuit breakers that are rated for high current dc. Fuses and breakers rated for ac will not work correctly with dc circuits. They will fail when you need them the most.

The breaker that feeds my inverter is a dc-rated 175 A job. It has terminals to allow up to AWG 0000 wires to be connected. It's a brute, and well worth the $350 price tag. Yup! That's correct. The price of the DC-250 disconnect that holds the breaker and bus bars is that expensive. That is cheap considering the job it is doing, protecting my house from a fire caused by a short circuit on the battery wiring that supplies the inverter.

USE A STANDARD COLOR CODE AND STICK WITH IT

While there's nothing carved into stone about what wires should be what color in a solar electric system, some standards have emerged. Not everyone uses these, but I recommend that you do.

For dc systems:
Black is negative.
Red is positive to the battery.
Yellow is positive to the photovoltaic (PV) array.
Green is ground.

For ac systems:
Black is hot.
White is neutral.
Green is safety ground.
Bare copper is safety ground.

Do you see now why you should never have both flavors in the same box?

Use Underwriters Laboratory (UL) rated or listed components.

Make sure the wire and connectors all have a UL listing or rating. A lot of charge controllers do not have a UL rating. That does not make them bad or inferior. Getting a UL listing is very expensive and time consuming. A lot of smaller companies do not have the funds for a UL listing or rating. Likewise a CE rating from the European Union is very expensive to obtain and requires tons of paperwork.

You do want to make sure the wire and fuses carry a UL rating. Sometimes those cheap ATC-style fuses you see in the surplus houses are made overseas and who knows their quality? I sure don't want to burn my house down by trying to save a buck on cheap fuses.

The wire going to your battery bank should be UL rated as well. This means the AWG 0000 weld cables are not going to be sufficient. Yes, I use weld cable and have never had a problem with it. An electrical inspector will not pass your installation if you use weld cable for battery wiring and interconnecting cables.

DOCUMENTS

Get a copy of the *National Electrical Code* book. The part that will be of most interest is section 690. This covers solar electric and backup electrical systems. I can tell you that most electrical inspectors will follow these guidelines to the letter, as most have never dealt with low-voltage, high-current installations.

Armed with a copy of the *National Electrical Code*, you should also document all of the wiring you have installed in your system. No, you're not going to remember what you did a year ago, let alone four years from now. Write it down, and keep the files up to date. Also include when you added water to your batteries and how much water was required. Good record keeping is important.

Make notes that indicate when and what amperage fuse may have blown. Label all the circuit breakers and fuse holders. Have instructions printed and placed near transfer disconnect switches. Have extra fuses of the correct value on hand. Make some sort of operating instructions so someone can operate and troubleshoot your systems when you are not around.

PLAY "WHAT IF?"

While you can't predict every type of failure your system may develop, sometimes a good case of "what if?" can go a long way. Try to put yourself in the worst-case scenario. What could happen to your system that could cause it to fail in a way that you or someone else's safety is compromised?

SAFETY GUIDELINES

While it's hard to put down a list of dos and don'ts, here are some common sense items I've used when working with low-voltage, high-current battery-based systems.

- Use the proper tools for the job. An adjustable wrench works great at tightening up bolts on a bike, but not for battery terminals.
- Purchase the correct size open end or box end wrench that will fit the hardware on your battery. Mark this wrench as "battery only" and keep it in your toolbox just for battery hardware. Shorter-length tools work best. (They are less likely to touch the other terminal if the wrench slips.)
- Use "plastic dip" or layers of electrical tape on the ends of your battery tools. This way an accidental touching of terminals won't melt the tool.
- Keep a mild solution of baking soda and water nearby, just in case you get acid on your skin.
- Have a source of fresh flowing water on hand. Water flushing of the eyes is the best method of removing acid.
- Use a full-face shield when working on batteries.
- No open flames or sparks near the battery bank. This is especially true when the battery bank is charging.
- No smoking or open flames when refueling generators.
- GFI breakers should be used when operating portable and using a generator. That is also true when you use dc to ac inverters.
- A small 300 W inverter can kill you just as dead as the 120 V ac mains will.
- Never trust safety interlocks or disconnect switches. Turn the switch off then check with a voltmeter.
- Disable internal combustion engines by disconnecting the spark plug wire before refueling or working on the engine.
- Keep all guards in place, especially when rotating parts are involved.
- Do not keep large amounts of gasoline or diesel fuel in storage.
- Never place propane tanks in buildings, sheds, or — heaven forbid — inside your home.
- Some lead-acid batteries are quite heavy. Don't try to move

them unless you have the necessary equipment or help.

PERSONAL PROTECTIVE EQUIPMENT

Purchase and use any special personal protective equipment you may need. Besides safety glasses you must have a good commercial fire extinguisher. Don't get one of those cheap under-the-kitchen-cabinet extinguishers. You want a good 10-pound dry-chemical extinguisher with the correct rating for the type of fire you may have. If you have a gasoline-powered generator, then the fire extinguisher must be rated for liquid-fueled fires.

Alternatively, purchase a fire extinguisher that is rated for all fire types. These are known as ABC extinguishers. That way it will work on paper, electrical fires and liquid-fueled fires.

Don't forget your safety glasses or full-face shield. Flying battery acid or molten metal will just ruin a perfectly good pair of eyes. Safety glasses are cheap and there's no reason not to wear them when working on batteries and other high current circuits.

Use the correct gloves, too. While I sure don't like working on live circuits, sometimes you just can't avoid it. So, if you need to wear gloves, make sure they are rated for the job you intend to do. If you work with batteries, then you need a pair that will withstand the battery acid.

TAG OUT AND LOCK OUT

That's a term that may not be familiar to a lot of you. If you have ever worked in an industrial setting you know what I mean. Tag out and lock out means just that. You lock out the controls and then tag the lock with your name and reason for locking out the switch, control or whatever it may be. Look at this way. Let's say you're working on a bad breaker in the 120 V ac breaker box at Field Day. You stop the generator and proceed to fix the breaker. Joe ham drops by and notices the generator is stopped. Thinking it may have run out of gas, he fills the tank and pulls the cord. The generator starts up and quickly fries your butt. Had you tagged and locked out the generator three things would have happened.

1. Joe could not start the generator because your lock was in the starting mechanism.

2. Joe would have seen why the generator was not running.

3. You would have survived this year's Field Day.

I've got almost 35 years in the steel industry. All employees are issued their own personal safety lock. Should the need arise, I can lock out and tag any piece of equipment I operate. That way, I won't end up with nine fingers because someone pressed the start button.

Don't *ever* think of cutting off a safety lock on any piece of gear.

If you are working on a windmill tower, then your ground crew must be wearing hard hats. A half-inch wrench dropped from 90 feet can just ruin your day if it hits you on

Don't do stupid stuff!

the head. In fact if you're working on a rooftop, then everyone below you should have some sort of protective headgear on, too. No, ball caps won't cut it. Use an approved hard hat.

KNOW WHEN TO SAY NO

If you've gone to a hamfest lately, did you take a look around you? As a group, there are a lot of fat bald old guys walking around with handheld radios in their hands. Right now, can you pick up a five-gallon can of gasoline and carry it across a football field? I'd guess not. Well, then don't try it. You may only make it half way, and right on the 50 yard line fall over dead from a heart attack.

Of course it need not be from a full tank of gas. If you can no longer pick up the generator and put it on and off of the pickup truck, then don't. Batteries are heavy. Don't kill yourself moving them into the basement.

Likewise, if you're not into climbing towers, then don't. Get someone that will go up the windmill tower with a grease gun.

GET A GOOD FIRST AID KIT

Instead of giving you a list of stuff to put into a first aid kit, I recommend that you just spend the money and get a good one. Not a cheap $5 kit with a few aspirins and a band-aid or two, but a real honest-to-goodness first aid kit. One should set you back about $40 to $60. After you dropped several grand for solar panels, you can afford a quality first aid kit.

While we're speaking about first aid kits, if you don't know basic first aid procedures having a room full of first aid kits won't do anyone any good. Everyone, and I mean everyone, in your house should attend a first aid class. Laying there on the floor turning several shades of blue while your wife thumbs through a booklet trying to figure out what to do is deadly.

CPR

It's a life saver. Know how to give CPR. The classes are often free, and they are available everywhere. Take a CPR class and save a life.

THE LAST WORD ON SAFETY

Let me remind you what I told you earlier in this chapter: Don't do stupid things!

When I got old enough to push the lawn mower my old man told me, "I see you have grown used to having your fingers. Good. When the mower's grass discharge gets clogged up, don't put your hands in there to clean it out. Use a stick."

Well, like, duh! Let me see: a 3.5 Hp engine turning a sharpened high carbon steel blade spinning 3600 rpm. Yea, right. I' *not* going to stick my hand in there.

Well move forward in time several decades and guess what? Lawn mower manufactures had to design a safety device to stop that rotating sharpened steel blade in a second or two and stall the engine, because people were doing stupid things, like sticking their hands in the grass discharge.

Don't be like other people. Think before you put yourself in harm's way.

With low-voltage, high-current dc applications, you should never work on a live circuit. Even though it's only 12 V dc, there can be enough current to cause some serious damage to tools and other metal objects if they get across a terminal.

Don't just jump in. Use your voltmeter to verify that the circuit is dead. Then check it again. Use insulated tools if you must work on live circuits. Never defeat circuit breakers and fuses. They are there to protect the wiring and the load. Did I mention you can burn down your house just as quickly on 12 V dc as you can with 120 V ac?

Planning and operating your own emergency power system can be a lot of fun. It's up to you to make it safe for you, your family and anyone else that may come in contact with it.

Safety doesn't just happen. It is planned. Make sure you plan it into your system from day one.

Know when to say NO!

Chapter 12

Emergency Practices

This last chapter deals not so much with generating or using emergency power, but how to keep you and your family safe. This chapter was also the hardest to write. It's hard to write about something you've never done. I have never had a member of the National Guard pound on my door telling me I have ten minutes to pack and get out, I can't relate to those that had to leave just like that.

I did a lot of digging, talked to a lot of different people and added a good dose of common sense. I hope some of the information presented in this chapter may help you and your loved ones if and when you get that knock at the door.

FIRST THINGS FIRST

Although the August 2003 blackout happened without notice, some disasters do come with some warnings. If you're under a thunderstorm warning, then there's a good chance you may see a tornado. Likewise, if you live along the eastern seaboard, hurricanes will be your biggest problem. When a hurricane warning is issued, that is the time to get serious.

Our technology has reached the point where we can predict most hurricanes and tornadoes and the area they will impact. While we're not to the point of narrowing the path of a tornado to street level, forecasting is getting better all the time.

I mention this because there is not always enough lead time to get your gear together before the storm hits. Let's look at some of the things you will need.

LONG TERM PLANNING

Many disasters are weather related. Even the 2003 blackout was in some ways caused by hot weather. The hot weather caused a lot of strain on the wires due to power usage and then the wires touched some trees. While we can't predict every disaster that comes — and many do come unannounced — a little bit of preplanning can go a long way. Since we kind of know ahead of time that things may go wrong, now is the time to do some planning. Get your ducks in a row, so to speak.

It simply amazes me when I watch CNN coverage of a hurricane. There is always video of hoards of people buying plywood at five times the normal price. Why would you wait until the hurricane warning flags are flapping in the wind to get plywood to cover your windows? To my way of thinking, that should be done in January instead of the middle of hurricane season.

Likewise, if you have an emergency generator, the generator should be started every few weeks, or at least once a month. That way, you know it will start when it's needed. Again, I've seen CNN footage of people trying in vain to start their backup generators as the flood waters slap against the house.

So do a bit of very early planning. Get the necessary equipment and make sure it works. Now, plan some more.

The first thing on the list is to make up two emergency evacuation routes. If you have to leave your home, you need to know how to get out of Dodge.

You need at least two evacuation routes to at least two different locations. Don't rely on the expressways. They may very well end up being a 35-mile long parking lot. Once you have your routes planned, tell family and friends where you plan to be in case an earthquake, hurricane or whatever disaster stops by to say hello.

Get to know the owner or manager of the local gas station, convenience store or mom and pop store. It is always nice to be friendly with people. Having a friendly, open relationship with the store owner can benefit both of you. Offer to adjust the SWR on his son's CB antenna. Look at his broken fish finder for him. Do the woman a favor. Be their friend! When the lights go out, you're out of cash, and the ATM machines won't work, you might be able to cash a check with the owners. They might be able to hold your check until the banking system gets back up. He might even loan you the needed supplies until the banking system is back up. You need a friend.

MONEY

Although it is hard at times to get both ends to wave at each other let alone meet, you need to keep some cash on hand. A hundred dollars or so in small bills will do. A roll or two of quarters, nickels and dimes is a good thing to have as well. This money needs to be set aside and not touched. It's for emergency use only.

No large bills, just ones, fives and tens. Why no large bills? Well, if you have ever sold anything at a hamfest or garage sale I am sure this has happened to you. After the necessary price haggling, the buyer gets the seller down to his lowest price. Then the buyer opens his wallet and pulls out enough cash to buy and sell most small countries. That can really be upsetting.

So just imagine you need some gasoline for the generator. You might be able to talk a local farmer into selling you some. And then you're going to pay him with a $100 bill? Guess how much a gallon of gas will be tomorrow! How is he going to make change? No, pay him in small bills. Better yet, offer to keep some of his frozen food in your generator-powered freezer. Barter what you have for what you need.

There are a lot of bad people around. When safety services become overloaded, the bad people spring up like weeds in a garden. So, don't flash your money. Don't mention to anyone that you have cash on your person. While it is true that during the darkest periods, people's kindness abounds, it is also true that those same conditions bring out the worst in others. Don't be a fool. Keep only a few bucks in your wallet. Use small bills.

FLASHLIGHTS, BATTERIES AND RADIOS

Always keep a working flashlight or two with extra batteries and bulbs nearby. Keep the batteries in the refrigerator. Do not put them in the freezer. Rotate the stock and try to keep fresh batteries at home at all times. Don't forget a spare flashlight bulb either. They do burn out.

Order a battery pack for your handheld radio that will accept "AA" batteries. You may not be in a location where you can recharge your NiMH battery pack. Take your wall wart charger with you, though. You may be able to plug into a socket and recharge. If not, then the AA battery holder will let you at least listen in on the public service frequencies. Program your handheld with those public service frequencies ahead of time.

Should you take a cell phone? Sure. Why not? Just remember that during most emergencies the cell phone towers are overcome with traffic. Most likely you won't be able to get service. Don't forget the phone's charger.

I have a small portable radio that receives shortwave, standard broadcast band and the FM broadcast band. It will operate on batteries or with a hand-operated generator. It's not the best shortwave radio made, but it works good enough for the intended application. I paid about $40 for this radio at a local RadioShack.

KEEP THINGS QUIET

If you have to go to a shelter, then by all means bring along a set of earphones or headphones. There's nothing quite like hearing a radio blasting away all day long to get someone's dander up. While you may be absolutely enthralled in what is going on by listening to the local police and fire, that elderly couple might have the bejeezus scared out of them. Now just think about that for a minute. Over your radio there are reports of massive fires breaking out in the old heights area, and that might very well be where the elderly couple call home. So, turn down the volume or use headphones when you're in a shelter.

PERSONAL ITEMS

I fully understand the price of prescription drugs. I also understand that most drugs have a short shelf life. So, your best bet is to have two or more places where you can renew your prescriptions if need be. Pick a drug store ten to twenty or more miles away. If you choose a national chain such as Walgreens, all their stores are linked by computer network. (That's assuming the link is running and there is power to operate the stores.)

For example, I get most of my prescription drugs from a local store that is about four miles away. I also have the ability to get drugs for my wife and myself at a Walgreens 21 miles away.

You have to do all of this beforehand. I took in a prescription for my wife and myself, registered with the store and got my name in the computer. Now, if need be, I can walk in and get a refill. This Walgreens is open 24/7 365 days a year, so if the time comes I have another supply for my prescription drugs.

WATER AND FOOD

The Red Cross says you need to plan on one gallon of water per person per day. If you have a well and a backup power system that has the ability to operate the well, then that's great. If not, you need to have a supply of water on hand.

Bottled water is by far the easiest to obtain. The trouble is, even bottled water has a shelf life. It's also expensive. You just can't afford to keep a gallon of water for every person per day in your house.

So, what to do? The Red Cross and FEMA say the next best thing is to fill the bathtub up with water. A few drops of chlorine bleach per gallon will kill most bugs and critters in the water. It may not taste great, but when it is the only drinkable water around, it will do. There are special filters and kits available that will make some of the worst water found anyplace drinkable. If you suspect your water supply maybe cut off or contaminated, then think about getting one of these kits. They're not cheap, but neither is water in the desert. Most backpack camping suppliers carry a variety of personal water filters.

Don't overlook water stored in your water heater tank. There's a good chance you have 40 to 80 gallons of drinkable water in the basement. You'll need a small hose to get the water out of the bottom drain valve on the water heater.

You can go only a few days without water, and a week without food, but at some time you need to eat. You need a supply of canned foods. Get foods that you can open and eat without adding water or milk. Keep a stock of foods that are naturally full of water. Canned pineapple and peaches provide both nutrition and fluids. They will also reduce the amount of water you need per day.

Of course, you will need a nonelectric can opener! Get several, they're only a few bucks each, just in case you break one or misplace it. Have you ever tried to open a can of soup without a can opener?

Frozen food will usually keep in the freezer for several days, provided you don't keep opening the door to check on

Emergency Practices

We have heard a lot talk about "go kits" and "jump Kits" that focus on getting communications up and running as quickly as possible. Have you ever given a bit of thought to a jump kit that just might well save your butt?

Most of my kit is shown laid out in the photo. Here is a breakdown of what is in the kit, as well as a quick explanation of why it's in there.

The first and primary reason the items are in this kit is to give you a better chance of staying alive for up to five days. The items shown are from my survival kit, and your kit will need to reflect the needs of your family and yourself.

I also suggest you duplicate your kit and its contents. It's best to have several on hand. There is truth in the old saying about putting all your eggs in one basket.

Let's take a look at the contents of my *stay alive kit*.

The first thing you may notice is the *lack* of communications equipment. When it comes time for me to leave Dodge, I want water and food. The radios can stay at home. That said, I do have two radios in the kit.

The first one is a Yaesu VX6. This multipurpose radio will cover the public service bands, TV audio, shortwave, and standard broadcast bands. Of course it will also function as a transceiver on 2 m and 70 cm.

The other radio is a Grundig FR200 recycle power radio. This radio covers AM, FM and two of the most popular shortwave bands. There's a built-in flashlight, too. But the best thing going for this radio is the fact it will run off of its own internal dynamo. You crank this baby up! Spinning the crank for several minutes will get you about half an hour or so of run time for the radio. The light is really rough on the battery. So if you plan on lighting up the area, be ready for some cranking. This radio (and light) will operate from a wall wart, the internal dynamo or from three AA batteries.

While not the best radio ever made, it does what it is designed to do. Finding a particular frequency on the shortwave bands is not easy, but it is possible. Oh yes, one more thing. It won't decipher CW or SSB. It's AM only on the shortwave bands.

For lighting, I have two sources; an LED light that uses one AA battery and a Pelican Saber light. The Pelican light produces a bright beam of light without draining the two C cells. Run time can be measured in days. The C cells will be much easier to obtain when everyone else is looking for D flashlight batteries! The LED light also produces a bright white light. It's designed for up close use while the Pelican can shine its beam for several hundred feet.

A one watt solar panel and a compact NiCd/NiMh battery charger rounds out the electronics. While the panel only produces one watt, it is one watt more than the person next to me probably has.

The WB8VGE "Keep Your Butt Alive Kit."

Health

You never know when you will have to go, so two rolls of toilet paper are a must. With strange food and uncertain water sources, a package of anti-diarrhea medication is a must as well. Make sure to include a supply of anti-bacterial cream. With all the stress and pulled muscles you're bound to have, a large bottle of Aspirin or other over-the-counter pain reliever is also a must have. I keep a handful of disposal gloves in my kit, too.

A small first aid kit is a required item! I picked this one up at Wal Mart for under $10. Just because they always work, throw in several books of matches. I included some paper money in small bills, and some coins.

By this time you're more than likely wondering why I included the cigarettes. Well first off, I don't smoke! Marlboro cigarettes are as good as cash (sometimes better) in just about any place on this planet! "I'll trade you a pack of Marlboros for two bottles of water."

Those are all the important things I plan to take with me if I need to go in a hurry. There are a few items that are not shown. You may want to include these in your kit was well.

• A six-foot length of ½" PVC hose for obtaining gasoline when no power is available.

• A small mini Maglite that uses AA batteries.

• Some chemical light sticks.

• A few simple hand tools such as screwdrivers and slip joint pliers.

• Extra batteries for the light.

• An assortment of "energy bars" for food.

• Clip leads and some insulated wire.

• Water purification kit.

What you don't see is any important papers. I keep those in a safety deposit box. I take my social security card, driver's license and car keys!

Everything in this kit will pack into one Craftsman nylon tool bag with room to spare. A small backpack would work, too. Make up a kit for everyone in your household. Then put them someplace where you can grab them quick, without having to think about where they are. Pray you'll never have to use them.

it. Avoid using ice, as the melting ice can contaminate the food. If a frozen foodstuff thaws, then use it or pitch it. Never refreeze any food that has thawed.

I have seen prepackaged dinners that do not need refrigeration. They have a shelf life of about a year, so they're great to have in your stock.

Dry foods are perfect to keep on hand, too, provided you have the necessary fluids to make them edible. Milk may be hard to come by, so don't plan on fixing a box of mac & cheese without it.

Yes, with the price of food on the rise, it's hard to purchase foodstuff and then have to pitch it when the expiration date is reached. I try to rotate the foodstuff into our meals. I'll take almost expired dinners to work or use them for one of our meals at home.

I also try to keep on hand at least two weeks of some sort of foodstuff at all times. They may not be the best tasting dinners I'll ever cook, but my wife and I won't go hungry either.

Try stocking energy bars. They have a very long shelf life and don't require refrigeration. They can, of course, also be eaten without heating them up.

If you really want to get crazy, you can always keep some military "meals ready to eat" (MRE) dinners. You can get these from several sources on the Internet. Look under MRE or survival Web sites. I have been told some of them are rather tasty. Some you would feel guilty feeding to the family dog.

HOT MEALS

Unless you really enjoy cold food, you need a way to heat your meals. There are a lot of choices. If you have natural gas or propane, then your gas range top will still work even if the power is out. You say you have a pilot-less ignition? No problem, use a match just like in the old days. Light match, then turn on gas. See Chapter 11 on safety, though. Don't do stupid things.

Next up is the gas BBQ grill. A grill with a side burner is great for heating up a can of soup. So it pays to have a full tank of propane or an extra full tank for grill cooking. If you don't have a gas grill, then a camper's stove would be great. A Coleman stove, complete with fuel will provide you and your family with hot meals. That is, provided you have extra fuel and a safe place to store this fuel.

Keep on hand some "canned heat" Sterno® fuel. This stuff keeps forever, and one can provide enough fuel to heat a complete meal for several adults.

KNOW WHERE YOU ARE

That may sound rather dumb. When the wild fires took out entire blocks of homes out West, the fires not only destroyed homes, but landmarks as well. Without landmarks, it was almost impossible to find what remained of your home. Street signs? Nope, they were melted. So the only way to know where you are was to know before you left. Either borrow or buy a GPS receiver and mark the location of your home. Write down the longitude and latitude of your home. Then keep several copies of this information, but not at the same place.

SANITATION

What goes in has got to come out. Sanitation is a must and most people don't give it much thought. I have a septic tank, so I don't have to worry about pipes to a city sewage treatment plant. All I need is water to flush the toilet. That may be a problem with limited electrical power. My water comes from my deep well pump. While I can operate the pump from my solar backup systems, there is always the possibility I may run out of power or have an inverter failure.

Depending on your situation, water for sanitation use may become more important than drinking water!

Depending on your location and type of sewer being used, you can get by with nothing more than a five gallon bucket and some water. No, you don't use the bucket instead of the toilet. The bucket is to hold water for flushing the toilet. If water is hard to come by, then by all means you don't want to use your only supply of potable water to flush down waste. Rainwater, water from melted snow or even water from a creek or drainage ditch will work. If you have to walk a half a mile to the creek to get water to flush the toilet remember, "If it's yellow let it mellow. If it's brown flush it down."

If you've never tried flushing a toilet without flipping the handle, it takes a bit of practice. First fill the five gallon bucket slightly more than half way. Then as fast as possible power the water into the toilet bowl. That's the key. If you have not figured it out yet, there's no need to flip the lever! You must pour the water in very quickly. If you don't, you will end up with a mess. Do *not* pour the water into the tank on the back.

IT'S YOUR TIME TO LEAVE

The tides are rising, the train has derailed and cats and dogs are sleeping together. Someone is knocking on your door telling you it's time for you get out. You've got ten minutes to pack your stuff and leave. If you want to stay, you need to fill out this form so we can contact your next of kin. What do you do?

You go, that's what. You may have read or heard about keeping a *jump kit* on hand. I know that isn't completely practical for 90% of us. If you live in an area prone to earthquakes or hurricanes, however, you should have one ready. They don't call these jump kits out West. They're known as personal survival packs. They contain energy bars for food, some money, change for pay phones, candy bars, water bottles, maps, small flashlights with extra batteries and other goodies.

What do you need to put in your survival kit? That depends on many factors, such as the length of time you may have to be away from home, your physical condition and your personal needs.

Most jump kits (not necessarily survival kits) are designed to be used at emergency Amateur Radio stations. They are not normally used to keep you alive. That's the difference between going to the emergency and one that is

Water

When hurricane Katrina kicked around New Orleans in August 2005, the mayor of the city told those that stayed behind they should bring enough food and water for upwards of five days. People were also told they should meet at the only shelter up and running — the Louisiana Superdome.

Now, let me see here. One gallon of water per person per day. My wife and I make two, so I need two gallons of water per day. And I will need enough for five days. Well that's ten gallons of water! At about eight pounds per gallon, that's 80 pounds to carry! The bottom line here is, you're just not going to be able to carry that much water with you. There has got to be an easier way to get water to drink, and there is. In fact, you have several avenues to take in search of drinkable water.

Chemical Disinfection

Chemical disinfection of water depends on the killing of bacteria, Giardia and amoeba cysts, and viruses by the chemical. Halogens (chlorine and iodine) are most commonly used. The important points are that the killing effectiveness of the chemical is dependant on concentration of the chemical, temperature of the water, and contact time. Decreased concentration (better flavor) or decreased temperature requires a longer contact time for disinfection. Sediment (cloudy water) increases the need for halogen. Bear in mind that adding flavor crystals to your water will use up the halogen and should only be done *after* the recommended contact time for disinfection. Remember: "Add flavor later."

There are several household chemicals that you can use to treat water. The first is chlorine. Chlorine has been used to disinfect water for several centuries. The most common objection to it is the flavor. If you live in the country and visit the city, you'll know first hand the taste of chlorinated water. There have been some suggestions that chlorine is unreliable in killing Giardia cysts in the commonly used concentrations.

Halizone Tablets

These are convenient and inexpensive, but have several disadvantages. Due to its chemical formulation, reliable disinfection in all conditions requires 6 tablets per liter for 1 hour of contact time. That results in poor flavor. The tablets rapidly lose effectiveness when exposed to warm, humid air. Halizone tablets would be hard to keep in good condition in the hot humid conditions of New Orleans.

Superchlorination-Dechlorination

This two-step method is somewhat inconvenient, and the chemicals needed are destructive to clothing and gear if spilled. It is highly effective and results in nearly flavorless water, though. High concentrations of chlorine are initially developed, and then in a second step removed by the addition of peroxide. Most people can't do superchlorination-dechlorination in a plastic Coke bottle. You need a water purification kit that will do the chemistry for you.

Iodine

Iodine has been used to disinfect water for nearly a century. It has advantages over chlorine in conve-nience and probably efficacy; many travelers find the taste less offensive as well. It appears safe for short and intermediate length use (3 to 6 months), but questions remain about its safety in long-term usage. **It should not be used by persons with allergy to iodine, persons with active thyroid disease, or pregnant women.**

Note that Iodine and other halogens appear to be relatively ineffective at killing cyclospora, a troublesome diarrhea-causing bacteria. It may be reasonable to pre-filter water to remove the large cyclospora (about the size of Giardia cysts), and then treat with iodine.

Iodine is available in numerous forms, which can be confusing to the user. These dosages work with one liter of water:

Iodine Topical Solution	2%	8 drops
Iodine Tincture	2%	8 drops
Lugol's Solution	5%	4 drops
Povidone-Iodine (Betadine®)	10%	4 drops
Tetraglycine hydroperiodide (Globaline®, Potable Aqua®, EDWGT®)	8 mg	1 tablet

If the water is clear and about 15°C let the iodine set for a minimum of 15 minutes

If the water is cloudy and about 15°C let the iodine set for a minimum of 30 minutes.

If the water is cooler, double the above times.

Filtering the Water

Filters work by physically removing infectious agents from the water. The organisms vary tremendously in size, from large parasitic cysts (*Giardia* and *Entamoeba histolytica* 5 to 30 μm), to smaller bacteria (*E. coli* 0.5×3 μm, *Campylobacter* 0.2×2 μm), to the smallest viruses (0.03 μm). Thus, how well filters work depends to a great extent on the physical size of the pores in the filter medium.

Filters have the advantage of providing immediate access to drinking water without adding an unpleasant taste. They suffer from several disadvantages, however; micro cracks or eroded channels within the filter may allow passage of unfiltered water, they can become contaminated, and no filters sold for field use are fine enough to remove virus particles (Hepatitis A, rotavirus, Norwalk virus, poliovirus, and others). In addition, they are expensive and bulky compared to iodine. Alas, many cheap filters are inadequate even to reliably remove *E. coli*, the most common infectious contaminant. Since your life may depend on getting a supply of water, don't go the cheap route when looking for a water filter. They all come with disposable filters and list what critters they will filter out.

They will also state how much water they will produce. Get one that will filter enough water for your needs.

The best bet is to use iodine to kill the critters and then a filter to remove the big stuff. A really good filter for emergency or portable use is the Potable Aqua Portable Water Purification Travel Kit. Check out the specs at: **www.safetycentral.com/potaqporwatp.html**.

pounding on your door at one in the morning.

So, what do you need to round up in those ten minutes before the bus comes?

Gather up all your prescription medicines. Bring along your personal ID such as a driver's license or stated-issued ID. Most of us should also carry a copy of our ham license, too.

Take your car keys! Your home maybe ruined and your car might have a cracked windshield, but it might also be drivable. If you leave your keys, you'll never find them. So, take your car keys with you. This applies even if you don't have to leave your home. A tornado might toss your house into the next county, but leave the family Buick untouched. If you left your keys in the house, you better hitch a ride to the next county and start looking for them.

On your way out, grab that roll of quarters. You can use coinage to work vending machines. Some vending machines don't require power; they're hand cranked! (Drop in your money, and turn the handle until your choice drops.) Remember, there is a good chance that ATM machines will be either empty or not working.

Turn off the gas going to your home. Pull the main service breaker. Everyone in your home should be able to do this. Everyone should know where the gas valve is and where the main electrical service box is.

TEDDY BEARS!

Yes, that sounds like a lifesaver. But really, did you know that many police, fire and highway patrol cars carry Teddy bears? When things go bad, there is something magical about holding on to a Teddy bear. It's almost like a drug. They're not just for young children, either. Adults respond just as well. I keep a few on hand, and my wife collects them. If and when I need to get up and go, some of them will go with me, not so much for me, but for that small child sitting beside me in the shelter whose life has been instantly uprooted. Allstate Insurance hands out about 12,000 Teddy bears a year to disaster victims.

CATS, DOGS AND CRITTERS

Most of the people that I have talked to say most shelters won't allow pets of any kind (except for service dogs). You can't take Spot or Fluffy with you to the shelter. So now what?

You leave them home. That's the general feeling most of the people that know have told me. You fill up the bathtub with water, leave out several bowls of water and plenty of food. Provided the house remains intact, the animals will survive.

Of course, you should have your pets' shots up to date and dogs should wear tags with your name and address stamped on them. Likewise if you cat wears a tag, it should identify you and not the cat. Smaller critters should be given ample food and water.

Now don't for a minute think that I don't care for animals. That's just not true. I have in my household six cats and two dogs. If I am told to get out in ten minutes, they all will have to stay behind. You just can't take them to the shelter. God willing, they will be safe until my wife and I return home.

HUNKERING DOWN AND WAITING OUT THE STORM

Sometimes, if you are well prepared, you can ride out the emergency. Of course there may not be any shelter you can get to, so you have to stay put. One example I can think of is a statewide ice storm like the one that hit parts of New England and Canada several years ago. The electrical wires were down, but most homes and business were not damaged. Of course there was no electrical power, no phones, no motor cars, not a single luxury. (Sorry, I couldn't help the *Gilligan's Island* reference.)

If you have used most of the information this book has to offer, you should now have electrical power, water and food. Of course you also have communications.

There is still a problem or two, though. Remember what I said about light in the darkness? Well if you have the only house for miles around with lights on, you're going to have friends like a dog has fleas. They'll be coming out of the woodwork. There are two avenues of thought here.

First, pool your resources. Your next-door neighbor may have an extra full propane tank, while you have power and lights. The next family over may have a freezer full of food. It's up to you to decide what and if you want to share resources. Keep in mind that you may have limited electrical backup. You more than likely don't have the power to keep your station on the air and supply the family with the power to run their freezer at the same time.

You might decide to just go it alone. You are completely off the grid and on emergency power. Use your own water and your own food.

Are you old enough to remember the 1962 Cuban missile crisis? If you do, then you may also recall Americans took to building "fallout shelters" by the thousands. How many were actually built? Nobody really knows because the best fallout shelter was the one that no one knew about. So, you may consider keeping quiet about what you have. That brings up another point.

FIREARMS

Remember the age-old question, "If you had to choose between the gold and the guns which one do you choose?" You would be surprised by the number of people that go with the gold. The correct answer is to choose the guns. For he that has the guns, also has the gold!

The gun issue in this country is a hot button. Both sides of the table have valid and compelling arguments. Do I own a gun? No. That being said, if I lived out in God's country I would have enough firepower to impress Dirty Harry. Firearms must be a personal issue only you can resolve. But, if you go with firearms, then make sure you have the training on their safe handling and operation. The National Rifle Association sponsors gun training and safety classes all over the country. Pop by their Web site for more information: **www.nra.org**.

RADIOS

The whole idea of this book is to keep the power flowing, enabling you to stay on the air. Now that you're sitting there in front of your radio equipment, do you have the training to really do anything helpful? I would suggest that you give traffic handling a try. The ARRL has volumes of material available to help you in generating and relaying messages. You should check into some of the many traffic handling nets to get acquainted with the hows and whys of traffic net operation. If nothing else, just sit back and listen to a traffic net session. Get to know the lingo and how the net operates. Then join in!

Not into HF traffic handling? Well neither am I, but that's not to say I don't get active in the local SKYWARN nets. Weather plays an important role in causing things to go kaput. So, join your local SKYWARN group. Know the ins and outs of traffic handling on a local level. Know how to report into net operation before trouble hits.

Don't overlook the Simulated Emergency Tests (SET) either. They are designed to bring out the best and the worst of a simulated emergency. Bring out your generator system, your backup batteries and whatever you plan on using when the real one pounds on your door.

Don't forget ARRL Field Day. While just about all of us have decided that Field Day is a contest, don't lose track that its main purpose is to provide a learning exercise in establishing portable emergency communications while out in the field. As a matter of fact, the ARRL has a category that allows you to operate at home using emergency power. Next spring, how about pulling the main breaker and see just how well everything meshes? Basically, camp at home. See if your lighting system is up to the 24 hour task. Try feeding the troops without using the stove. Just give it a try for 24 hours and you'll come away with a totally different outlook.

GET OTHERS INVOLVED

Show your husband, wife, girlfriend, boyfriend, daughter, son, mother or whomever how to turn on and operate some of your radio equipment. My first choice would be the 2-meter radios. If something would happen to me during an emergency and there's no one around, I know my wife could turn on the radio, push the microphone and get help. Legal? No, but if I am lying on the floor having a heart attack, we'll worry about the legal problems later.

My current emergency backup system is normally set for automatic operation and is transparent to everyone in the household. Just in case it doesn't switch over automatically, my wife also knows how to turn on the backup power system if need be. She also knows how to reset the circuit breakers and what loads will work when we are "on batteries."

THE JOURNEY IS OVER

We have come to the end of our journey of self-reliance and emergency power systems. I hope you have gained enough information to make logical, safe decisions for your family and your ham radio station.

Every situation is different and thus each one of you will need to design your backup system to match your needs. It's not a case of one size fits all.

Here is a case in point. I have always enjoyed driving and owning Jeep vehicles. A stock Jeep right off the showroom floor can go just about anyplace it is pointed. For some, that's not enough. They modify, add and change things to the point the modified Jeep can crawl over rocks the size of Volkswagen Beetles! Your emergency backup system will be the same way. A simple system will take you just about anyplace, or you can start climbing over rocks. You have the final say.

I have had a lot of fun designing and using my emergency power system. My system will never be *done* and I keep fine tuning and tinkering with it all the time. I'll bet you will be the same way with your system. The next time someone forgets to trim tree limbs under the power cables in Ohio, I'll see you on the HF bands running emergency power designed by ham radio operators. While the road may have ended, the adventure has just begun.

Appendix A

Wire Sizes for Connecting Solar Panels to Emergency Power Systems

15 Specification Voltage is 15.00 V dc

1 Wiring is specified for a power efficiency of 97.5% and/or correct ampacity

PV ARRAY CURRENT IN AMPERES.

		6	12	15	18	21	24	30	36	42	48	54	60	66	72	78	84	90	96	102	108	114	120
R	25	13	10	10	9	8	7	6	6	5	4	4	4	3	3	2	2	2	1	1	0	0	0
O	30	13	10	9	8	7	7	6	5	4	4	3	3	2	2	2	1	1	1	0	0	0	0
U	35	12	9	8	7	7	6	5	4	4	3	3	2	2	1	1	1	0	0	0	0	-1	-1
N	40	11	8	7	7	6	5	4	4	3	2	2	1	1	1	0	0	0	-1	-1	-1	-1	-1
D	45	11	8	7	6	6	5	4	3	3	2	1	1	1	0	0	0	-1	-1	-1	-2	-2	-2
	50	10	7	7	6	5	4	4	3	2	1	1	1	0	0	-1	-1	-1	-1	-2	-2	-2	-2
T	60	10	7	6	5	4	4	3	2	1	1	0	0	-1	-1	-1	-2	-2	-2	-3	-3	-3	-3
R	70	9	6	5	4	4	3	2	1	1	0	0	-1	-1	-2	-2	-2	-3	-3	-3	-3		
I	80	8	5	4	4	3	2	1	1	0	-1	-1	-1	-2	-2	-3	-3	-3					
P	90	8	5	4	3	3	2	1	0	0	-1	-2	-2	-2	-3	-3	-3						
	100	7	4	4	3	2	1	1	0	-1	-1	-2	-2	-3	-3								
W	125	7	4	3	2	1	1	0	-1	-2	-2	-3	-3										
I	150	6	3	2	1	0	0	-1	-2	-3	-3												
R	175	5	2	1	0	0	-1	-2	-3	-3													
E	200	4	1	1	0	-1	-1	-2	-3														
	225	4	1	0	-1	-1	-2	-3															
L	250	4	1	0	-1	-2	-2	-3															
E	275	3	0	-1	-2	-2	-3																
N	300	3	0	-1	-2	-3	-3																
G	325	2	-1	-2	-2	-0																	
T	350	2	-1	-2	-3	-3																	
H	375	2	-1	-2	-3																		
	400	1	-1	-2	-3																		
in	425	1	-2	-3																			
	450	1	-2	-3																			
F	475	1	-2	-3																			
E	500	1	-2	-3																			
E	600	0	-3																				
T	700	-1																					

Specification Voltage 15 V dc
for 12 V dc PV Applications

Wiring is specified for a power efficiency of 97.5%
and/or correct ampacity

Codes

The body of the table contains the Wire Gauge Number

"0" Wire is designated by 0

"00" Wire is designated by -1

"000" Wire is designated by -2

"0000" Wire is designated by -3

Wiring power efficiency is specified at 68°F.

If ambient temperature is > 90°F, use the next gauge larger wire

30   Specification Voltage is 30.00 V dc

1   Wiring is specified for a power efficiency of 97.5% and/or correct ampacity

PV ARRAY CURRENT IN AMPERES.

LENGTH	6	12	15	18	21	24	30	36	42	48	54	60	66	72	78	84	90	96	102	108	114	120
25	14	13	12	12	10	8	6	6	6	6	6	6	4	4	4	2	2	2	2	0	0	0
30	14	13	12	11	10	8	6	6	6	6	6	6	4	4	4	2	2	2	2	0	0	0
35	14	12	11	10	10	8	6	6	6	6	6	5	4	4	4	2	2	2	2	0	0	0
40	14	11	10	10	9	8	6	6	6	5	5	4	4	4	3	2	2	2	2	0	0	0
45	14	11	10	9	9	8	6	6	6	5	4	4	4	3	3	2	2	2	2	0	0	0
50	13	10	10	9	8	7	6	6	5	4	4	4	3	3	2	2	2	1	1	0	0	0
60	13	10	9	8	7	7	6	5	4	4	3	3	2	2	2	1	1	1	0	0	0	0
70	12	9	8	7	7	6	5	4	4	3	3	2	2	1	1	1	0	0	0	0	-1	-1
80	11	8	7	7	6	5	4	4	3	2	2	1	1	1	0	0	0	-1	-1	-1	-1	-1
90	11	8	7	6	6	5	4	3	3	2	1	1	1	0	0	0	-1	-1	-1	-2	-2	-2
100	10	7	7	6	5	4	4	3	2	1	1	1	0	0	-1	-1	-1	-1	-2	-2	-2	-2
125	10	7	6	5	4	4	3	2	1	1	0	0	-1	-1	-2	-2	-2	-2	-3	-3	-3	-3
150	9	6	5	4	3	3	2	1	0	0	-1	-1	-2	-2	-2	-3	-3	-3				
175	8	5	4	3	3	2	1	0	0	-1	-1	-2	-2	-3	-3	-3						
200	7	4	4	3	2	1	1	0	-1	-1	-2	-2	-3	-3								
225	7	4	3	2	2	1	0	-1	-1	-2	-3	-3	-3									
250	7	4	3	2	1	1	0	-1	-2	-2	-3	-3										
275	6	3	2	1	1	0	-1	-2	-2	-3	-3											
300	6	3	2	1	0	0	-1	-2	-3	-3												
325	5	2	1	1	0	-1	-2	-2	-3													
350	5	2	1	0	0	-1	-2	-3	-3													
375	5	2	1	0	-1	-1	-2	-3														
400	4	1	1	0	-1	-1	-2	-3														
425	4	1	0	-1	-1	-2	-3															
450	4	1	0	-1	-1	-2	-3															
475	4	1	0	-1	-2	-2	-3															
500	4	1	0	-1	-2	-2	-3															
600	3	0	-1	-2	-3	-3																
700	2	-1	-2	-3	-3																	
800	1	-1	-2	-3																		

(Left vertical label: ROUND TRIP WIRE LENGTH in FEET)

Specification Voltage 30 V dc
for 24 V dc PV Applications

Wiring is specified for a power efficiency of 97.5%
and/or correct ampacity

Codes
The body of the table contains the Wire Gauge Number
"0" Wire is designated by 0
"00" Wire is designated by -1
"000" Wire is designated by -2
"0000" Wire is designated by -3

Wiring power efficiency is specified at 68°F.
If ambient temperature is > 90°F, use the next gauge larger wire

Installing Anderson Powerpole® Connectors

ARES and RACES groups have generally adopted the use of Anderson Powerpole® connectors for power supply and radio connections. By using these standard power connectors, any radio can plug into any power source that uses the Anderson Powerpole® connectors.

While the standard is good for everyone, some hams are having a hard time working with Powerpoles®. There are several important steps you must take to assemble a set of Powerpole® connectors. Before we get much further, we need to know some Anderson Powerpole® lingo.

THE ANDERSON POWERPOLES®

The Powerpoles® comes in various sizes and capacities. The ones you and I use are rated at 15, 30 or 45 A. The capacity rating is based on the contact. If you want a Powerpole® with a current rating of 30 A, then you must purchase the 30 A contact.

The contact lives in the housing. Using either the 15, 30 or 45 A contacts, you need housings PP15/45. All three contact types are the same physical size, and fit in the same housing. The PP15/45 housing accommodates any one of the contacts listed above.

SELECT THE PROPER CONTACT

So, what contact do you use? That depends on a few factors. One of course is the amount of current you want to use. If you're planning on wiring up a small dc to ac inverter, then 45 A contacts would be required. On the other hand if you're going to put a set of Powerpoles® on the power lead for a 2 m radio, then 15 A contacts would be just fine.

Needless to say, you can't get 10-gauge wire into a 15 A contact. Trying to crimp a 16-gauge wire into a 45 A contact

won't work either. So, the bottom line on contact selection is:
• Select the contact for the required current.
• Select the contact that will fit correctly on the end of the wire.
• Select and assemble the housing pair.

Now that we have the contact problem solved, it's time to look at those housings again. For 99% of us, we use red and black housings. I use yellow for solar panel connections and gray for any battery voltage above 32 V. That's my personal choice.

When you pick up a Powerpole® connector, you will notice that the housing has slots and grooves molded into each side. The housings will mate with each other by slipping one into another of those grooves. Be aware that there is an almost unlimited number of combinations for the ori-

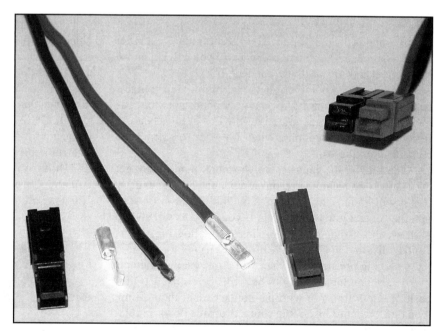

Figure 1 — Here is the proper way of setting up the housings. A pair of connectors is ready to be installed on one end of the wire and another pair has already been installed on the other end.

entation of the connector bodies to slide together. So, to keep everyone on the same page, the ARES/RACES standard is, red for positive, black for negative.

When you assemble the pair, orient them so the red connector is towards the right and the "tongue" in the connector is to the top. "Red right, tongue top" is the standard orientation that ARES/RACES has adopted.

Put the red housing in your right hand with the part of the housing that the wire goes in away from you. Now rotate the housing so the tongue is at the top and you're good. Then hold the black housing to the left of the red one, in the same orientation. See **Figure 1**.

Are you confused? It's like trying to describe to someone in writing how to unpeel and eat a banana without them knowing what a banana looks like. Try it this way.

When viewed from the *contact end* (opposite the wire side), the tongues inside the housing go up, the hood goes down, red on the right and black on the left.

This is the ARES/RACES standard way to assemble these connectors! One of the interesting things about these connectors is that two pairs of wires with the connectors installed this way will plug into each other. There is no *male* and *female* end to the connector block.

ADDING CONTACTS TO THE WIRES AND CRIMPING THE CONTACTS

This is the hardest part when it comes to assembling Powerpoles®. Getting the contacts on the wires without mangling the contacts can be a challenge. I've assembled a lot of Powerpoles®, and have found a few tricks. Here's how to do it, painlessly!

Make sure the wire and the contact match. Wire that is too large won't work with small contacts. Ditto for small wire gauge and too-large contacts.

Use the proper tool to crimp the wires onto the contacts. Yes, you can get by with pliers and other crimpers, but for the best crimp, use the proper contact tool. The best one that I have found is the one supplied by West Mountain. It's not cheap — about $50 — but use it once and you will never use anything else to install these connectors.

Decide how you will crimp the wires onto the contacts *before* you crimp them! This is a very important step. Almost all the problems people have with Powerpoles® can be traced to improper planning when it comes to getting the contacts onto the wire.

The heavier the gauge cable, the more important to get the contacts on right. Install the contact on the wire exactly like it will go into the housing. You don't want to twist or jam the contact into the housing. If you do, you will end up with a contact that will distort inside the housing. If the contact distorts, you will have problems trying to plug into the matching housing. So when you crimp on the contacts make sure you have them on the wires correctly. I find it easier to do a trial run with the contacts, and then crimp. You don't want to install the contacts upside down in relationship to the housings. When you have the contacts on the wire the right way, they will go into the housing without trouble.

It's easy to get the contact on one way and then put the other contact on upside down. Trying to twist and jam the contact into the housing won't work. When done correctly, the contacts with wires attached will simply slide right into the housing.

CRIMPING OR SOLDERING?

I have tried them both and have found if you have the proper crimping tool, you really don't need to solder the contacts. If you don't crimp the contacts correctly, they won't fit the housing. You need the proper crimp tool.

Likewise, if you do a sloppy soldering job, the contacts won't fit in the housing either. A large glob of solder will prevent the contact from seating in the housing. If you get the contacts too hot, then they won't work correctly either. They'll loose their spring and fail to mate with their counterpart in the matching housing. You never want to get solder on the face of the contact.

I've solder and crimped contacts, and I have just solderd and just crimped. I believe if you have the proper size wire and the correct contact, and the right crimping tool, there is no need to solder the wire and the contact.

INSTALLING THE CONTACTS INTO THE HOUSING

The contacts go in only one way. Insert the contacts with their sharp, curved edge down against the flat spring that is in the housing. They should slide in and click. If you do not hear a click or they are not fully seated, fix them. When they are inserted fully you should notice that the contact and it's wire "floats" slightly inside the housing. If it feels tight it may not be snapped in fully or you have made the contact wider than it originally was during crimping or soldering

A FEW ODDS AND ENDS

Sometimes when you get a set of Powerpoles® they come with little steel roll pins. *Do not use the roll pins.* They will fall out and drop inside someplace they should not be, usually at the most inopportune time. If you have trouble keeping the housing together, a drop of super glue will work just fine. *Do not use the roll pins!*

Don't disconnect the connectors while there is current in the circuit. The contacts and housings are not designed to interrupt current. Damage will result to the Powerpoles®.

Don't pull on the wires when trying to separate the housing. Disconnect a Powerpole® just like you would a 120 V ac plug from the wall. You pull the plug out by the

Although I was using contacts rated at 45 A, this pair melted under high current. The plastic housings merged into one.

To prevent melting connectors, I upgraded all my heavy-current connections to Anderson Powerpole® SB connectors. Unlike the PP45, these can only mate with like-colored connectors. A yellow connector will not fit inside a gray connector. I installed contacts that will accept 6-gauge wire and will handle upwards of 125 A.

plug and do not yank it out with the cord.

The Powerpoles® do not have a great deal of mechanical resistance when it comes to holding one housing to another[1]. That being said, don't mount the Powerpole® connectors upside down. Between gravity and the weight of the wires, they will pull the connections apart. Ditto on using them in a mobile application. If you need to make sure the connectors do not sepa-

The SB series connector is left of the ARES/RACES 45 A connector. Quite a bit of difference!

[1]Anderson Power products does make a high-detent contact set that has much more holding power. These contacts are special order. I don't know if you can use standard detent contact with high detent contacts. Contact Anderson Power Products.

rate when in use, take a Nylon cable tie and lace one around both connector housings.

Make up adapter cables that have one end terminated in Powerpole® connectors. That way you can still connect to a power source even if the other guy has not converted. I have a cable made up that has a cigar plug that goes into any automobile, with Powerpole® connectors on the other end. Likewise, I don't cut off those goofy power plugs on the ends of the power cables that come with my radios. I just make up a Powerpole® jumper wire that will plug into the radio's connector and terminate with a pair of Powerpoles® on the other end.

When you buy these connectors, buy a lot of them. You will be surprised by the number of radios and what-nots that can benefit from Anderson Powerpoles®.

If you belong to ARES/RACES, using this standard power connector is a must. Even if you don't belong to ARES/RACES, I suggest you convert all of your radio equipment over to Powerpoles®.

Organize a Powerpoles® connector party for your local radio club. Have all the club members come with their radios or power cables and crimp the night away. If the club has enough members that would be interested in this, perhaps the club could purchase a pair of crimpers. Purchase your Powerpole® contacts and housings in bulk for the best prices.

Now that the standard has been set, you should join the club. When the next disaster hits, knowing that all of your radios will plug into a power source with Anderson Powerpoles® is one less thing to worry about.

Appendix C

Selected Emergency Power Projects from *QST*

In the pages the follow you will find a collection of helpful articles selected from the pages of *QST*. These articles cover a variety of topics that supplement the detailed information included in this book. There are accessories that you will find helpful for use in your emergency power system and ideas about how others have implemented emergency power for their Amateur Radio stations.

A Low-Voltage Disconnect

Whether you're operating a repeater, operating on emergency power or just watching TV in your RV, this little gadget protects your batteries from damage.

By Michael Bryce, WB8VGE

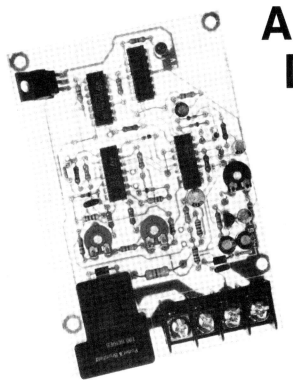

M any Amateur Radio stations use batteries to power their radio equipment during commercial electric power outages. Some of us use battery power all the time in the shack. Keeping an eye on the battery's charge sometimes can't be done (or is forgotten altogether) until you are unexpectedly— and unwillingly—off the air!

What you need is a battery watchdog— something to keep track of the battery and disconnect loads when the battery just about goes kaput. No matter what your use of battery power—whether you own a camper, RV, or just fish on the lake beside your cottage—this contraption does the battery monitoring for you and protects your battery from severe discharge as well. Repeater owners and operators may find the device an ideal way to extend operating time while on emergency power.

What's It Do?

The low-voltage disconnect (LVD) automatically disconnects a load from the battery before damage is done to the battery or the load. The potential across a discharging battery's terminals depends on the battery's state of charge and the load's discharge rate. This circuit monitors the battery terminal voltage, and when it reaches a preset level, a relay is de-energized disconnecting the load from the battery. There's an approximate 5-second delay before the device senses the low-voltage set point and the relay drops.

Take a look at some of the features of this watchdog:

- User-adjustable turn-off voltage.
- User-adjustable reset voltage.
- Built-in delay to ignore temporary low-voltage conditions.
- 30-A-capacity relay contacts.
- Low current consumption.
- Easy construction using a readily available PC board and components.

The load can be connected directly to the LVD's relay, or its relay can drive an off-board, heavy-duty power relay, if need be. You can also use the PC-board-mounted relay to control a 120-V ac load (I'll talk about that later). The LVD can also supply logic to a repeater controller, too.

Circuit Description

To see how the circuit works, refer to Fig 1. U5, an LM317LZ 100-mA adjustable-voltage regulator, creates a reference voltage for the comparators. This voltage (4.00) is set by R30 (REFERENCE ADJ), a 1-kΩ potentiometer. R27 places a 4-mA load on the regulator's output and improves regulator stability. U2C buffers the reference voltage. From here, the reference voltage goes to the voltage comparators. D1, a 1N4001 diode, protects U5 from reversed power-supply voltage polarity.

R1 and R2 halve the battery terminal voltage. R3 and C9 help filter out battery-line noise. U2B acts as a buffer between the voltage divider and the battery sense line.

Two set points are needed to control the LVD. One turns on the LVD. That causes K1 to drop out, disconnecting the load. The other set point turns off the LVD, closing the relay contacts and reconnecting the load. If it weren't for the two different set points, the LVD would constantly switch on and off. The difference in the set-point voltages allows the battery to recover before the load is reconnected. (This assumes you have some means of recharging the battery after the load has been disconnected.)

The voltage comparators are nearly identical. The battery voltage, now divided by two, is applied to two voltage dividers. Let's first look at the trip comparator.

The battery's output voltage is divided in half and applied to comparators U1C and U1B through U2B. If the battery was discharged to 10 volts,[1] the trip comparator (U1C) input would see less than 5 volts. The reference voltage at pin 10 is 4 volts. By adjusting R5 (**LVD TRIP**), we can set the comparator to switch states when the input voltage at U1C pin 9 equals the reference voltage. D5 and R12 provide a bit of hysteresis to keep U1C from oscillating. Because U1C may not provide the needed high-to-low positive switching action, a second op-amp section (U1D) is used. U1D provides the switch-like on/off state needed by the delay circuit composed of D4, R16 and C8, which provides a delay of about 5 seconds. Again, to provide the required logic levels, U2A is

[1]Notes appear at the end of this article.

used. Its output goes to the SET point of the R/S latch, U4A and B.

The **ON RESET** circuit (U1B) works similarly, but has no delay circuit. U1B's output is routed through U1A and goes to the RESET of the R/S latch, U4A.

When the battery voltage is above the LVD trip point, the output of the R/S latch (U4A pin 3) is high. This turns on Q1, a TIP-120 NPN Darlington power transistor, which energizes K1. A 1-W, 47-Ω resistor (R18) limits relay current. This minimizes the overall current demand of the LVD. D8 protects Q1 from the back EMF produced when the relay coil's magnetic field collapses.

As the battery discharges, its terminal voltage falls. When the LVD turn-on voltage is reached, there's a 5-second delay, then the output of U2A goes low, setting the R/S latch. This removes Q1's base drive, causing K1 to de-energize. The output of U4 pin 4 connects to U4 pin 8 to allow the oscillator (U2D) to output through U4C to U3A, which turns on the **LOW-VOLTAGE** LED, DS1. DS1 flashes at a rate determined by R25, R26 and C7. With the values shown, the on-time is about 1/20th of a second. Otherwise, DS1 remains dark. This arrangement reduces the circuit current drain during a low-voltage battery condition.

Construction

There's nothing critical here: perfboard, wire-wrap, dead-bug or PC-board construction are all suitable. A PC board is available[2] Using a PC board speeds construction and makes troubleshooting easier.

You can buy most of the parts from a well-stocked Radio Shack store. Mouser Electronics[3] can supply the parts "the Shack" does not carry. In both cases, the single exception is the relay. Obviously, the one specified fits the PC board. Secondly, it has a hefty contact rating (30 A) and is inexpensive (less than $4). This relay is available from Digi-Key.[4] Certainly, you can use a different relay, but you'll probably have to mount it off-board. Also, you may have to change R18's value, as it's dependent on the relay's coil current. The value shown is calculated for the relay identified in the caption. (By the way, mount R18 with a $^{1}/_{4}$- to $^{1}/_{2}$-inch clearance between the resistor body and the PC board. This allows air to circulate around the resistor and prevents the dissipated heat from discoloring the PC board.)

Component values aren't critical. Use equivalent parts you have on hand. If you don't have a 300-kΩ resistor, a 270-kΩ

will work just fine. (The ±10% rule of thumb can be safely applied.) Use sockets for the ICs and be careful when handling U4: It's subject to damage by static discharges, so use a wrist strap. Install all the parts—*except for C8*. After you've assembled the PC board, check your work to ensure the diodes and capacitors are properly polarized. Ensure you have the ICs oriented properly before you apply power to the board.

Set Up and Test

You'll need a digital VOM[5] and a variable-voltage power supply to adjust the LVD. Connect your power supply to the terminal block at the battery terminals.

Set the power supply output to 14 volts. Apply power to the board. Probe pin 10 of U2. Verify the presence of the reference voltage; in all probability, it's not 4.00 volts—yet. Move the probe to U2 pin 8 and verify the reference voltage is there, too. Now, adjust R30 (**REFERENCE ADJ**) until the voltage at U2 pin 8 is 4.00 volts. Next, set your power supply to 10.5 volts. Measure the voltage at pin 7 of U2. It should be 5.25 volts (half of 10.5 volts).

Adjustment Method

First, *remove C8* (in case you forgot the earlier warning and soldered it in) from the PC board. This defeats the delay circuit. Turn both R7 and R5 fully counterclockwise. Set your power supply output to 13.5 volts. Connect the power supply to the LVD, then turn it on. The **LOW VOLTAGE DISCONNECT** LED (DS1) should be blinking. Slowly adjust R7 (**ON RESET**) until the relay closes and DS1 goes dark. Reduce the power supply output to 10.5 volts. Slowly turn R5 (**LVD TRIP**) until K1 drops out and DS1 begins to flash. Verify the two set points by raising the power supply output voltage to 13.5 volts. K1 should energize. Reduce the voltage to 10.5 volts and K1 drops out. *Now* you can install C8! You're done!

Troubleshooting

If you can't get the circuit to work, first check for the presence of the reference voltage. Without it, you'll be dead in the water from the start. Check for 4.00 volts on pins 3, 6, 10 and 12 of U1, and at U2 pin 2 and U3 pin 3.

If the comparators won't switch (and you have the proper reference voltage) check the battery sense line by checking the output voltage at U2 pin 7. (This voltage should be one-half of the power-supply voltage applied to the battery sense

line.) As you can see, replacing U1 replaces *all* the battery sense comparators.

When the battery voltage is at 10.5, you'll be able to see the delay action by probing U2 pin 3. At this pin, you'll see the voltage slowly drop during the 5-second delay period.

If you used a relay other than the one specified, the value of the current-limiting resistor (R18) may be too high to allow the relay to energize. Try reducing R18's value (or short it out).

Hooking Up the LVD

With only four wires, hook-up is a breeze! Simply connect the battery you intend to monitor to TB1's battery terminals. K1's contacts are completely isolated from the battery. By connecting a jumper from the +12-volt battery terminal to one of the relay contacts, you can deliver battery power through the relay contacts to your load.

Remember, although K1's contacts can carry 30 amperes, voltage drop caused by long wire runs can have an effect on the load. If you need to control heavy current loads, use K1 to control a power relay located right at the load. Install a protective diode across the power-relay's coil terminals to prevent inductive kickback.

You can use K1's contacts to control logic levels to a repeater. Connect K1's contacts to ground or +12 V via a current-limiting resistor. If you have a repeater controller and it requires a logic input, this is one way to go.

Controlling a 120-V AC Load

Although K1's contacts easily handle a 120-volt load, having an exposed 120-volt line connected to TB1 would keep me up at night! A safer way to control such a load is to use an off-board solid-state relay (see Fig 2).[6] (A solid-state relay is an optically coupled device that provides excellent isolation between the load and the driving source. In this case, between the 120-V ac mains and your battery.) Solid-state relays with various control voltages are readily available. RX, an external resistor, serves to limit the current flowing to the solid-state relay. By properly altering the value of RX, you can use a 5- or 12-volt control line. As mentioned earlier, the battery can supply power to operate the solid-state relay via K1's relay contacts.

Life with an LVD

While the LVD certainly protects your battery from deep discharge, it's not perfect. (What is?) Every LVD consumes

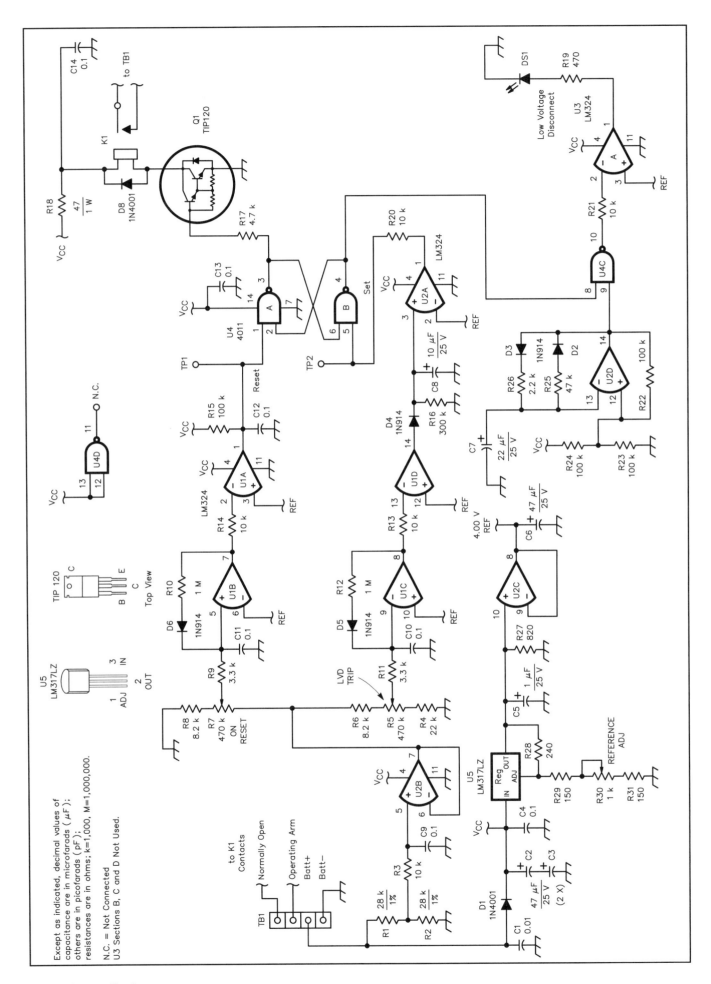

Fig 1—Schematic of the low-voltage disconnect circuit. Equivalent parts can be substituted. Unless otherwise specified, resistors are ¼-W, 5%-tolerance carbon-composition or film units.

U1-3—LM324 quad op amp (Mouser 511-LM324; Digi-Key LM324N).

U4—4011 quad gate (Mouser 511-4011; Digi-Key 4011CD).

U5—LM317LZ 5-V, 1-A adjustable regulator (Mouser LM317LZ; Digi-Key LM317LZ).

K1—Potter and Brumfield T-90 series; 12-V dc, 155-Ω coil, SPST, 30-A normally open contacts (Digi-Key PB110-ND).

TB1—Terminal block (Mouser 506-8PCV-04).

R5, R7—470-kΩ trimmer potentiometer (Mouser 531-PT15D-470K).

R30—1-kΩ trimmer potentiometer (Mouser 531-PT15D-1K).

Q1—TIP-120 Darlington power transistor (Mouser 511-TIP-120; Digi-Key TIP120PH-ND).

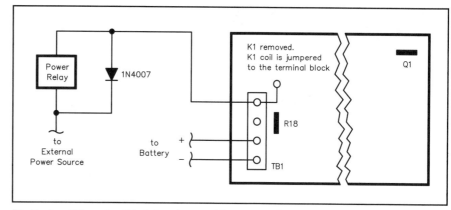

Fig 2—An off-board solid-state relay provides an excellent means of controlling 120-V-ac operated loads. Here, the relay derives its operating current and voltage from the battery being monitored. The resistance value and power rating of RX is chosen for proper relay operating current.

some power from the battery it's trying to protect. In this case, when the LVD has the relay pulled in, it draws about 90 mA. If you run the LVD 24 hours a day, you have a 2.16-Ah load just for the LVD. Even with the relay off, and the battery at 10.5 volts, the LVD draws about 12 mA. So, if your battery is being charged by a solar array, be sure to include the LVD load requirements when performing your sizing calculations. To save power when you're not using the load, turning off the LVD automatically turns off the load connected to the relay.

If your battery is charged from a 120-V ac charger, you'll not have to worry about the extra LVD load. Repeaters operators normally have a battery back-up system constantly being charged. When the grid power fails, the battery takes over. When the battery discharges to the point that the LVD trips, the LVD can then take the power amplifier off line to extend battery operation until the grid

power comes back on.

The LVD load should not be your main load. *Shedding* loads is the main job of the LVD. You don't want to have it shut down *everything,* but disconnect what you can live without. For instance, in an RV, you may want to connect the running lights to the LVD and leave your TV bypassed. When the battery becomes so low as to trip the LVD, the running lights will be disconnected. (Given the nature of what's on TV these days, it's probably a better idea to take the TV off line and keep the running lights on!)

Here's another example of choice: You have a sailboat docked in the lake. A bilge pump is connected to the LVD. If too much water leaks into the boat and the pump is running all the time, the LVD will disconnect the pump from the battery protecting the battery from damage. With the pump disconnected from the battery, the pump won't work any more and before you know it, your sailboat has become a submarine!

A much better way to prevent your sail-boat from sinking is to have the LVD warn you of the discharged battery. The warning could be as simple as a flashing

light or a buzzer. If you really want to go to the extreme, you could combine the LVD and the METCON II[7] for telemetry. The LVD keeps a constant eye on your batteries, while you work the world on your radio, or just fish in the lake by your cottage.

Notes

[1] A lead-acid battery is generally considered dead when the terminal voltage is 10.5 under load.

[2] A PC board for this project is available from FAR Circuits, 18N640 Field Ct, Dundee, IL 60118-9269. Price: $12, plus $1.50 shipping. A PC-board template package is available free from the ARRL. Address your request for the LOW VOLTAGE DISCONNECT TEMPLATE to: Technical Department Secretary, ARRL, 225 Main St, Newington, CT 06111. Please enclose a business-size SASE.

[3] Mouser Electronics, 2401 Hwy 287 N, Mansfield, TX 76062; tel 800-346-6873, 817-483-4422, fax: 817-483-0931.

[4] Digi-Key Corp, 701 Brooks Ave S, PO Box 677, Thief River Falls, MN 56701-0677, tel 800-344-4539, 218-681-6674, fax 218-681-3880.

[5] You can use an analog meter to calibrate the LVD, but a digital voltmeter provides better resolution.

[6] Available from All Electronics Corp, PO Box 567, Van Nuys, CA 91408-0567, tel 800-826-5432, 818-997-1806, fax 818-781-2653.

[7] P. Newland, "Introducing METCON, a New Remote control and Telemetry System," *QST,* Jan 1993, pp 41-47.

By Mike Bryce, WB8VGE

The Micro M+ Charge Controller

Current capacity of up to
4 A, positive line switching
so all grounds tie together,
standby current of less than
1 mA and more features
make the Micro M+ the ideal
photovoltaic charge controller for use at home or
in the field. It's an easy-to-build, one-evening project that just about anyone can master.

T he Micro M proved a very popular project.[1] It seems hams really do like to operate their rigs from solar power while in the outback. Many hams find solar power to be very addictive. I had dozens of requests for information on how to increase the current capacity of the original Micro M controller. The original Micro M would handle up to 2 A of current. The PC board traces and blocking diode limited the design to this current capacity. I also wanted to improve the performance of the Micro M while I was at it. Because the Micro M switched the negative lead of the solar panel on and off, the negative lead of the solar panel had to be insulated from the system ground. While that's not a problem with portable use, it may cause trouble with a home station where all the grounds should be connected. Here's what I wanted to do:

• Reduce the standby current at night

• Increase current handling capacity to 4 A

• Change the charging scheme to high (positive) side switching

• Improve the charging algorithm

• Keep the size as small as possible, but large enough to build.

The Micro M+

I called the end result the Micro M+. You can assemble one in about an hour. Everything mounts on one double-sided PC board. It's small enough to mount inside your rig yet large enough so you won't misplace it. You can stuff four of them in your shirt pocket! And, you need not worry about RFI being generated by the Micro M+. It's completely silent and makes absolutely zero RFI!

The Micro M+ will handle up to 4 A of current from a solar panel. That's equal to a 75-W solar panel.[2] I've reduced the standby current to

[1]Notes appear at end of article.

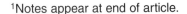

Figure 1—
This photo shows
the Micro M+ charge
controller circuit board. Leads
solder to the board and connect to a
solar panel and to the battery being charged.

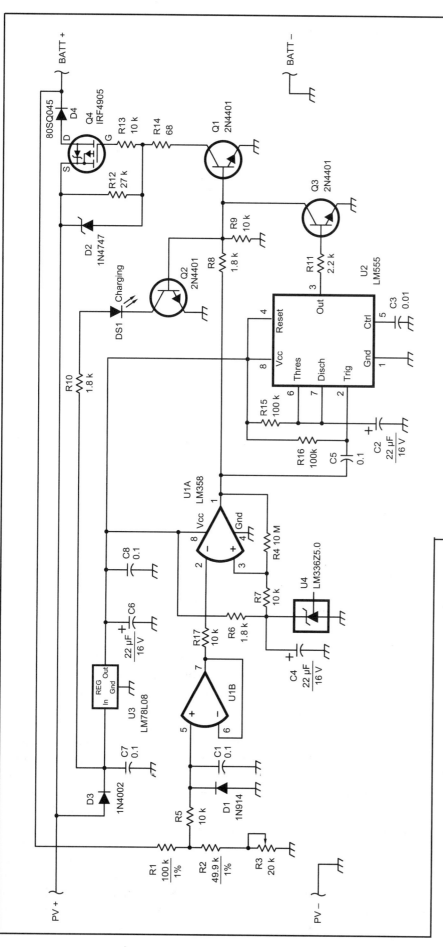

Figure 2—The schematic diagram of the Micro M+ charge controller.
C1, C5, C7, C8—0.1 µF.
C2, C4, C6—22-µF, 16-V electrolytic.
C3—0.01 µF.
D1—1N914, small signal silicon switching diode.
D2—1N4747, 20-V, 1-W Zener.
D3—1N4002, silicon rectifier diode.
D4—80SQ045, 45-V, 8-A Schottky diode.
DS1—LED, junkbox variety.
Q1, Q2, Q3—2N4401 NPN small-signal transistor (2N2222 or 2N3904 will also work).
Q4—IRF4905 P-channel MOSFET in TO-220 case. You will also need a small clip-on heat sink for this case.
R1—100 kΩ, 1%.
R2—49.9 kΩ, 1%.
R3—20-kΩ trimmer.
U1—LM358AN, dual op-amp.
U2—LM555AN timer.
U3—LM78L08, 8-V regulator.
U4—LM336Z-5.0, 5.0-V Zener diode in TO-92 case. The adjust terminal allows control of the temperature coefficient and voltage over a range. The adjust terminal is not used for the Micro M+.

less than 1 mA. I've also introduced a brand new charging algorithm to the Micro M+. All the current switching is done on the positive side. Now, you can connect the photovoltaic (PV) array, battery and load grounds together.

A complete kit of parts is available as well as just the PC board. The complete kit, including the PC board and all parts is $30.[3] The Micro M+ is easy to build, making it a perfect first-time project.

Here's How it Works

Figure 1 shows the complete Micro M+, while Figure 2 shows the schematic diagram. Let's begin with the current handling part of the Micro M+. Current from the solar panel is controlled by a power MOSFET. Instead of using a common N-channel MOSFET, however, the

Micro M+ uses an International Rectifier IRF4905 P-channel MOSFET. This P-channel FET has a current rating of 64 A with an RDS_{on} of 0.02 Ω. It comes in a TO-220 case. Current from the solar panel is routed directly to the MOSFET source lead.

N-channel power MOSFETs have very low RDS_{on} and even lower prices. To switch current on and off in a high side application, the gate of an N-channel MOSFET must be at least 10 volts higher than the rail it is switching. In a typical 12-volt system, the gate voltage must be at least 22 volts to ensure the MOSFET is turned completely on. If the gate voltage is less than that required to fully enhance the MOSFET, it will be almost on and somewhat off (the MOSFET is operating in its linear region). The device will be destroyed at high current.

To produce this higher gate voltage, some sort of oscillator typically is used to charge up a capacitor via a voltage doubler. This charge pump generates harmonics that may ride on the dc flowing into the battery under charge. Normally, this would not cause any problem, and in most cases, a filter or two on the dc bus will eliminate most of the harmonics generated. Even the best filter won't get rid of all the harmonics, however. To compound the problem, long wire runs to and from the solar panels and batteries act like antennas.

The P-channel MOSFET eliminates the need for a charge pump altogether. To turn on a P-channel MOSFET, all we have to do is pull the gate lead to ground! Since the Micro M+ does not have a charge pump, it generates *no RFI*!

Now, you may be wondering, if the P-channel MOSFET is so great, why have you not seen them in applications like this before? The answer is twofold. First, the RDS_{on} of a P-channel MOSFET has always been much higher than its N-channel cousin. Several years ago, a P-channel MOSFET with an RDS_{on} of 0.12 ohms was considered very low. At

that time an N-channel MOSFET had an RDS_{on} of 0.009 ohms. Suppose you want to control 10 A of current from your solar panel. Using the N-channel MOSFET above we find the MOSFET will dissipate less than a watt of power. On the other hand, the P-channel MOSFET will dissipate 12 W of power! Current generated by our solar panels is way too expensive to have 12 W of it go up as heat from the charge controller.

The second factor was price. The P-channel MOSFET I described above would have sold for $19 each. The N-channel would have been a few dollars.

The Micro M+ never draws current from the battery. The solar panel provides all the power the micro M+ needs.

In the last year or so the RDS_{on} of the P-channel MOSFET has fallen to 0.028 ohms. The price, while still a bit on the steep side, has dropped to about $8 each.

With the P-channel MOSFET controlling the current, diode D4—an 80SQ045 Schottky—prevents current from the battery from flowing into the solar panel at night. This diode also provides reverse polarity protection to the battery in the event you connect

the solar panel backwards. This protects the expensive P-channel MOSFET.

Zener diode D2, a 1N4747, protects the gate from damage due to spikes on the PV line. Resistor R12 pulls the gate up, ensuring the power MOSFET is off when it is supposed to be.

The Micro M+ Likes to Sleep

The Micro M+ never draws current from the battery. The solar panel provides all the power the Micro M+ needs. At night, the Micro M+ goes to sleep. When the sun rises, the Micro M+ will start up again. As soon as the solar panel is producing enough current and voltage to start charging the battery, the Micro M+ will pass current into the battery.

To reduce the amount of standby current, diode D3 passes current from the solar panel to U3, the voltage regulator. U3, an LM78L08 regulator, provides a steady + 8 V to the Micro M+ controller. Bypass capacitors C6, C7 and C8 are used to keep everything happy. As long as the solar panel is producing power, the Micro M+ will be awake. At sundown, the Micro M+ will go to sleep. Sleep current is on the order of less than 1 mA.

Using the Micro M+ with the Yaesu FT-817

With the introduction of the new Yaesu FT-817 all mode, all band QRP transceiver, more and more of us will be using solar power in the field. The Micro M+ was designed to use a 12-V solar panel to charge a 12-V battery. The Yaesu FT-817 can operate from 12 V supplied externally or from an internal 9.6-V NiCd battery. The NiCd battery may be charged when the battery is installed in the radio. Or, if you want, it can be charged separately from the 817 via a solar panel and the Micro M+ controller.

To use the Micro M+ to charge this NiCd pack, you'll have to change the value of resistor R2 from 49.9 kW 1% to 82.5 kW 1%. This will allow the logic to switch correctly at 11.6 V, the voltage of a fully charged 9.6-V NiCd battery. This assumes you use the standard of 1.45 V per NiCd cell. With the new value for R2, there's plenty of adjustment in the state-of-charge trimmer to allow you to fine-tune the state-of-charge.

Since the NiCd battery is rated at only 9.6 V, this throws the power point of the solar panel in the trash. A typical 5-W solar panel is rated at 290 mA at 17.1 V. Because of the lower battery voltage, there will be more than the 290 mA of current flowing. However, if the panel is designed to produce 5 W, that's all it will do. As the voltage goes down, the current will increase, up to the Isc (current short circuit) of the panel. The panel will not produce any more current than it was designed for.

Battery Sensing

The battery terminal voltage is divided down to a more usable level by resistors R1, R2 and R3. Resistor R3, a 20-kΩ trimmer, sets the state-of-charge for the Micro M+. A filter consisting of R5 and C1 helps keep the input clean from noise picked up by the wires to and from the solar panel. Diode D1 protects the input of the op-amp in the event the battery sense line were connected backward.

An LM358 dual op-amp is used in the Micro M+. One section, U1B, buffers the divided battery voltage before passing it along to the voltage comparator, U1A. Here the battery sense voltage is compared to the reference voltage supplied by U4. U4 is an LM336Z-5.0 precision diode. To prevent U1A from oscillating, a 10-MW resistor is used to eliminate any hysteresis.

As long as the battery under charge is below the reference point, the output of U1A will be high. This saturates transistors Q1 and Q2. Transistor Q2 conducts and lights LED DS1, our CHARGING LED. Q1, also fully saturated, pulls the gate of the P-channel MOSFET to ground. This effectively turns on the FET and current flows from the solar panel into the battery via D4.

As the battery begins to take up the charge, its terminal voltage will increase. When the battery reaches the state-of-charge set point, the output of U1A goes low. With Q1 and Q2 now off, the P-channel MOSFET is turned off, stopping all current into the battery. With Q2 off, the CHARGING LED goes dark.

Since we have basically eliminated any hysteresis in U1A, as soon as the current stops, the output of U1A pops back up high again. Why? Because the battery terminal voltage will fall back down as the charging current is removed. If left like this, the Micro M+ would sit and oscillate at the state-of-charge set point.

To prevent that from happening, an LM555 timer chip, U2, monitors the output of U1A. As soon as the output of U1A goes low, this low trips U2. The output of U2 goes high, fully saturating transistor Q3. With Q3 turned on, it pulls the base of Q1 and Q2 low. Since both Q2 and Q1 are now deprived of base current, they remain off.

With the values shown for R15 and C2, charging current is stopped for about four seconds after the state-of-charge has been reached.

After the four second delay, Q1 and Q2 are allowed to have base drive from U1A. This lights up the charging LED and allows Q4 to pass current once more to the battery.

As soon as the battery hits the state-of-charge once more, the process is repeated. As the battery becomes fully charged, the "on" time will shorten up while the "off" time will always remain the same four seconds. In effect, a pulse of current will be sent to the battery that will shorten over time. I call this charging algorithm "Pulse Time Modulation."

As a side benefit of the pulse time modulation, the Micro M+ won't go nuts if you put a large solar panel onto a small battery. The charging algorithm will always keep the off time at four seconds allowing the battery time to rest before being hit by higher current than normal for its capacity.

Building Your Own Micro M+

There's nothing special about the circuit. The use of a PC board makes the assembly of the Micro M+ quick and easy. It also makes it much easier if you need to troubleshoot the circuit. You can build the entire circuit on a piece of perf-board if you want.

The power MOSFET must be protected against static discharges. A dash of common sense and standard MOSFET handling procedures will work best. Don't handle the MOSFET until you need to install it in the circuit. A wrist strap would be a good idea to prevent static damage. Once installed in the PC board, the device is quite robust.

A small clip-on heat sink is used for the power MOSFET. If you desired, the MOSFET could be mounted to a metal chassis. If you do this, make sure you insulate the MOSFET tab from the chassis.

If you plan on using the Micro M+ outside, then consider soldering the IC directly onto the board. I've found that cheap solder-plated IC sockets corrode. If you want to use an IC socket, use one with gold-plated contacts.

Feel free to substitute part values. There's nothing really critical. I do suggest you stick with 1% resistors for both R1 and R2. This isn't so much for the close tolerance, but for the 50-PPM temperature compensation they have. You can use standard off-the-shelf parts for either or both R1 and R2, but the entire circuit should then be located in an environment with a stable temperature.

Adjustments

You'll need a good digital voltmeter and a variable power supply. Set the power supply to 14.3 V. Connect the battery negative and power supply negative leads together at a circuit-board ground point. Connect the PV positive and battery positive lead, and the power supply positive leads together. The charging LED should be on. If not, adjust trimmer R3 until it comes on. Check for +8 V at the V_{cc} pins of the LM358 and the LM555. You should also see + 5 V from the LM336Z5.0 diode.

Quickly move the trimmer from one end of its travel to the other. At one point the LED will go dark. This is the switch point. To verify that the "off pulse" is working, as soon as the LED goes dark quickly reverse the direction of the trimmer. The LED should remain off for several seconds and then come back on. If everything seems to be working, it's time to set the state-of-charge trimmer.

Now, slowly adjust the trimmer until the LED goes dark. You might want to try this adjustment more than once as the closer you get the comparator to switch at exactly 14.3 V, the more accurate the Micro M+ will be. Here's a hint I've learned after adjusting hundreds of Micro M+ controllers. Set the power supply to slightly above the cutoff voltage you want. If you want 14.3 V, then set the supply to 14.5 V. I've found that in the time it takes to react to the LED going dark, you overshoot the cutoff point. Setting the supply higher takes this into account and usually you can get the trimmer set to exactly what you need in one try. That's all you need to do. Disconnect the supply from the Micro M+ and you're ready for the solar panel.

Odds and Ends

The 14.3-V terminal voltage will be correct for just about all sealed and flooded cell lead-acid batteries. You can change the state-of-charge set point if you want to recharge NiCds or captive sealed lead-acid batteries.

Keep the current from the solar panel within reason for the size of the battery you're going to be using. If you have a 7-amp hour battery, then don't use a 75-W solar panel. You'll get much better results and smoother operation.

The tab of the power MOSFET is electrically hot. If you plan on using the Micro M+ without a protective case, make sure you insulate the tab from the heat sink. A misplaced wire touching the heat sink could cause real damage to both the Micro M+ and your equipment. A small plastic box from RadioShack works great.

More Current?

Well yes, you can get the Micro

The Micro M+ Charge Controller board, small enough to mount inside your rig, is shown connected to a solar panel and a rechargeable battery.

M+ to handle more current. You must increase the capacity of the blocking diode and mount the power MOSFET on a larger heat sink. I've used an MBR2025 diode and a large heat sink for the MOSFET and can easily control 12 A of current.

Battery Charging Without a Solar Panel?

Yes, that's possible, too. The trick is to use a power supply for which you can limit the output current. A discharged lead acid battery will draw all the current it can from the charging source. In a solar panel setup, if the panel produces 3 A, that's all it will do. With an ac powered supply, the current can be excessive. To use the Micro M+ with

an ac powered supply, set the voltage to 15.5 V. Then limit the current to 2 or 3 A.

No matter if you're camping in the outback, or storing photons just in case of an emergency, the Micro M+ will provide your battery with the fullest charge. The Micro M+ is simple to use and completely silent. Just like the sun!

Notes

[1]"The Micro M," Sep 1996 *QST*, p 41.
[2]A 75-W module produces 4.4 A at 17 V. The Micro M+ can easily handle the extra 400 mA.
[3]A complete kit of parts is available from SunLight Energy Systems, 955 Manchester Ave SW, North Lawrence, OH 44666. A complete kit including all parts and PC board is $30 plus $4 US Priority mail. Visa, MasterCard accepted. Tel 330-832-3114; **www.seslogic.com/**.

By Bob Lewis, AA4PB

An Automatic Sealed-Lead-Acid Battery Charger

This nifty charger is just what you need to keep your SLA batteries up to snuff!

Photos by Joe Bottiglieri, AA1GW

After experiencing premature failure of the battery in my Elecraft K2 transceiver (most likely because I forgot to keep the battery on a regular charge schedule), I began searching for an *automatic* battery charger.[1,2] The K2 uses a Power-Sonic PS-1229A 12-V, 2.9-Ah sealed lead-acid (SLA) battery. SLAs are commonly called *gel-cells* because of their gelled electrolyte. As with all things, to obtain maximum service life from an SLA battery, it needs to be treated with a certain degree of care. SLA batteries must be recharged on a regular basis; they should not be undercharged or overcharged. If an SLA battery is left unused, it will gradually self-discharge.

Although my SLA battery experiences related here are linked to my K2 transceiver, you can think of the K2 simply as a load for the battery. The comments pertaining to the SLA batteries and chargers apply across the board and the charger described here can be used with any similar battery.

Using a Three-Mode Charger

My first attempt at keeping my K2's SLA battery healthy was to purchase an automatic three-mode charger. I soon discovered that most three-mode chargers work by sensing

[1]Notes appear at end of article.

current and were never intended to charge a battery under load.

Three-mode chargers begin the battery charging process by applying a voltage to the battery through a 500-mA current limiter. This stage is known as *bulk-mode* charging. As the battery charges, its voltage begins to climb. When the battery voltage reaches 14.6 V, the charger maintains the voltage at that level and monitors the battery charging current. This is known as the *absorption mode*, sometimes called the *overcharge mode*. By this time, the battery has achieved 85% to 95% of its full charge. As the battery continues to charge—with the voltage held constant at 14.6 V—the charging current begins to drop. When the charging current falls to 30 mA, the three-mode charger switches to *float mode* and lowers the applied voltage to 13.8 V. At 13.8 V, the battery becomes self-limiting, drawing only enough current to offset its normal

self-discharge rate. This works greatæuntil you attach a light load to the battery, such as turning on the K2 receiver. The K2 receiver normally draws about 220 mA. When the charger detects a load current above 30 mA, it's fooled into thinking that the battery needs charging, so it reverts to the absorption mode, applying 14.6 V to the battery. If left in this condition, the battery is overcharged, shortening its service life.

UC3906-IC Chargers

Chargers using the UC3906 SLA charge-controller IC work just like the three-mode charger described earlier except that their return from float mode to absorption mode is based on voltage rather than current. Typically, once the charger is in float mode it won't return to absorption mode until the battery voltage drops to 10% of the float-mode voltage (or about 12.4 V). Although this is an improvement over the three-mode

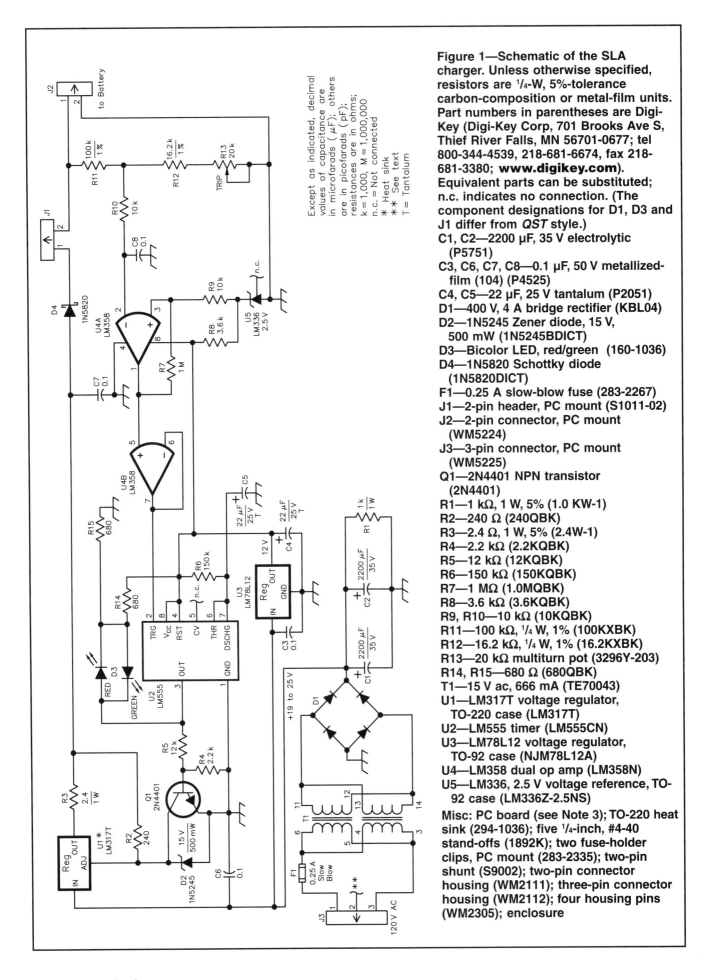

Figure 1—Schematic of the SLA charger. Unless otherwise specified, resistors are ¼-W, 5%-tolerance carbon-composition or metal-film units. Part numbers in parentheses are Digi-Key (Digi-Key Corp, 701 Brooks Ave S, Thief River Falls, MN 56701-0677; tel 800-344-4539, 218-681-6674, fax 218-681-3380; **www.digikey.com**). Equivalent parts can be substituted; n.c. indicates no connection. (The component designations for D1, D3 and J1 differ from *QST* style.)

C1, C2—2200 µF, 35 V electrolytic (P5751)
C3, C6, C7, C8—0.1 µF, 50 V metallized-film (104) (P4525)
C4, C5—22 µF, 25 V tantalum (P2051)
D1—400 V, 4 A bridge rectifier (KBL04)
D2—1N5245 Zener diode, 15 V, 500 mW (1N5245BDICT)
D3—Bicolor LED, red/green (160-1036)
D4—1N5820 Schottky diode (1N5820DICT)
F1—0.25 A slow-blow fuse (283-2267)
J1—2-pin header, PC mount (S1011-02)
J2—2-pin connector, PC mount (WM5224)
J3—3-pin connector, PC mount (WM5225)
Q1—2N4401 NPN transistor (2N4401)
R1—1 kΩ, 1 W, 5% (1.0 KW-1)
R2—240 Ω (240QBK)
R3—2.4 Ω, 1 W, 5% (2.4W-1)
R4—2.2 kΩ (2.2KQBK)
R5—12 kΩ (12KQBK)
R6—150 kΩ (150KQBK)
R7—1 MΩ (1.0MQBK)
R8—3.6 kΩ (3.6KQBK)
R9, R10—10 kΩ (10KQBK)
R11—100 kΩ, ¼ W, 1% (100KXBK)
R12—16.2 kΩ, ¼ W, 1% (16.2KXBK)
R13—20 kΩ multiturn pot (3296Y-203)
R14, R15—680 Ω (680QBK)
T1—15 V ac, 666 mA (TE70043)
U1—LM317T voltage regulator, TO-220 case (LM317T)
U2—LM555 timer (LM555CN)
U3—LM78L12 voltage regulator, TO-92 case (NJM78L12A)
U4—LM358 dual op amp (LM358N)
U5—LM336, 2.5 V voltage reference, TO-92 case (LM336Z-2.5NS)

Misc: PC board (see Note 3); TO-220 heat sink (294-1036); five ¼-inch, #4-40 stand-offs (1892K); two fuse-holder clips, PC mount (283-2335); two-pin shunt (S9002); two-pin connector housing (WM2111); three-pin connector housing (WM2112); four housing pins (WM2305); enclosure charger, it still has the potential for overcharging a battery to which a light load is attached.

First, let's look at the situation where a UC3906-controlled charger is in absorption mode and you turn on the K2 receiver, applying a load. The battery is fully charged, but because the load is drawing 220 mA, the charging current never drops to 30 mA and the charger remains in absorption mode, thinking that it is the battery that is asking for the current. As with the three-mode charger, the battery is subject to being overcharged.

If we remove the load by turning off the K2, the current demand drops below 30 mA and the charger switches to float mode (13.8 V). When the K2 is turned on again, because the charger is able to supply the 220 mA for the receiver, the battery voltage doesn't drop, so the charger stays in float mode and all is well. However, if the transmitter is keyed (increasing the current demand), the charger can't supply the required current, so it's taken from the battery and the battery voltage begins to drop. If we unkey the transmitter before the battery voltage reaches 12.4 V, the charger stays in float mode. Now it takes much longer for the charger to supply the battery with the power used during transmit than it would have if the charger had switched to absorption mode.

Let's key the transmitter again, but this time keep it keyed until the battery voltage drops below 12.4 V. At this point, the charger switches to the absorption mode. When we unkey the transmitter, we're back to the situation where the charger is locked in absorption mode until we turn off the receiver.

Why Worry?

So, why this concern about overcharging an SLA battery? At 13.8 V, the battery self-limits, drawing only enough current to offset its self-discharge rate (typically about 0.001 times the battery capacity, or 2.9 mA

for a 2.9 Ah battery). An SLA battery can be left in this float-charge condition indefinitely without overcharging it. At 14.6 V, the battery takes more current than it needs to offset the self-discharge. Under this condition, oxygen and hydrogen are generated faster than they can be recombined, so pressure inside the battery increases. Plastic-cased SLA batteries such as the PS-1229A have a one-way vent that opens at a couple of pounds per square inch pressure (PSI) and release the gases into the atmosphere. This results in drying the gelled electrolyte and shortening the battery's service life. Both undercharging and overcharging need to be avoided if we want to get maximum service life from the battery.

Continuing to apply 14.6 V to a 12-V SLA battery represents a relatively minor amount of overcharge and results in a gradual deterioration of the battery. Applying a potential of 16 V or excessive bulk-charging current to a small SLA battery from an uncontrolled solar panel can result in serious overcharging. Under these conditions, the overcharging can cause the battery to overheat, which causes it to draw more current and result in thermal runaway, a condition that can warp electrodes and render a battery useless in a few hours. To prevent thermal runaway, the maximum current and the maximum voltage need to be limited to the battery manufacturer's specifications.

Design Decision

To avoid the potential of overcharging a battery with an automatic charger locked up by the load, I decided to design my own charger, one that senses battery voltage rather than current in order to select the proper charging rate. A 500-mA current limiter sets the maximum bulk rate charge to protect the battery and the charger's internal power supply. Like the three-mode chargers, when a battery with a low terminal voltage is first connected to the charger, a con-

stant current of 500 mA flows to the battery. As the battery charges, its voltage begins to climb. When the battery voltage reaches 14.5 V, the charger switches off. With no charge current flowing to the battery, its voltage now begins to drop. When the current has been off for four seconds, the charger reads the battery voltage. If the potential is 13.8 V or less, the charger switches back on. If the voltage is still above 13.8 V, the charger waits until it drops to 13.8 V before turning on. The result is a series of 500-mA current pulses varying in width and duty cycle to provide an average current just high enough to maintain the battery in a fully charged condition. Because the repetition rate is very low (a maximum of one current pulse every four seconds) no RFI is generated that could be picked up by the K2 receiver. Because the K2's critical circuits are all well regulated, slowly cycling the battery voltage between 13.8 V and 14.5 V has no ill effects on the transmitted or received signals.

As the battery continues to charge, the pulses get narrower and the time between pulses increases (a lower duty cycle). Now when the K2 receiver is turned on and begins drawing 220 mA from the battery, the battery voltage drops more quickly so the pulses widen (the duty cycle increases) to supply a higher average current to the battery and make up for that taken by the receiver. When the K2 transmitter is keyed, it draws about 2 to 3 A from the battery. Because the charger is current limited to 500 mA, it is not able to keep up with the transmitter demands. The battery voltage drops and the charger supplies a constant 500 mA. The battery voltage continues to drop as it supplies the required transmit current. When the transmitter is unkeyed, the battery voltage again begins to rise as the charger replenishes the energy used during transmit. After a short time, (depending on how long the transmitter was keyed) the battery voltage reaches 14.5 V and the pulsing begins again. The charger is now fully automatic, maintaining the battery in a charged condition and adjusting to varying load conditions.

The great thing about this charging system is that during transmit the majority of the required 2 to 3 A is taken from the battery. When you switch back to receive, the charger is able to supply the 220 mA needed to run the receiver and deliver up to 280 mA to the battery to replenish what was used during transmit. This means that the power source need only supply the average energy used over time, rather than being required to supply the peak energy needed by the transmitter. (You don't need to carry a heavy 3-A regulated power supply with your K2.) As long as you don't transmit more than about 9% of the time, this system should be able to power a K2 indefinitely.

Have you ever noticed that sometimes when your H-T has a low battery and you drop it into its charger you hear hum on the received signals? This charger's power supply is well filtered to ensure that there is no ripple or ac hum to get into the K2 under low battery voltage conditions.

Circuit Description

The charger schematic is shown in Figure 1. I've dubbed the charger the PCR12-500A, short for Pulsed-Charge Regulator for 12-V SLA batteries with maximum bulk charge rates of 500 mA. U1, an LM317 three-terminal voltage regulator, is used as a current limiter, voltage regulator and charge-control switch. A 15-V Zener diode (D2) sets U1 to deliver a no-load output of 16.2 V. R3 sets U1 to limit the charging current to 500 mA. When Q1 is turned on by the LM555 timer (U2), the **ADJ** pin of U1 is pulled to ground, lowering its output voltage to 1.2 V. D4 effectively disconnects the battery by preventing battery current from flowing back into U1. A

Schottky diode is used at D4 because of its low voltage drop (0.4 V).

An LM358 (U4A) operates as a voltage comparator. U5, an LM336, provides a 2.5-V reference to the positive input (pin 3) of U4. R11, R12 and R13 function as a voltage divider to supply a portion of the battery voltage to pin 2 of U4A. R13 is adjusted so that when the battery terminal voltage reaches 14.5 V, the negative input of U4A rises slightly above the 2.5-V reference and its output switches from +12 V to 0 V. When this happens, the 1-MW resistor (R7) causes the reference voltage to drop a little and provide some hysteresis. The battery voltage must now drop to approximately 13.8 V before U4A turns back on.

U4B is a voltage follower. It pulls the trigger input (pin 2) of U2 to 0 V, causing its output to go to 12 V. U4B's output remains at 12 V until C5 has charged through R6 (approximately four seconds) *and* the trigger has been released by U4A sensing the battery dropping to 13.8 V or less. While the output of U2 is at 12 V, emitter/base current for Q1 flows via R5 and Q1's collector pulls U1's **ADJ** pin to ground, turning off the charging current.

The output of U2 also provides either +12 V or 0 V to the bicolor LED, D3. R14 and R15 form a voltage divider to provide a reference voltage to D3 such that D3 glows red when U2's output is +12 V and green when U2's output is at 0 V. When ac power is applied but U1 is switched off and not supplying current to the battery, D3 glows red. When U1 is on and supplying current to the battery, D3 is green. As the battery reaches full charge, D3 blinks green at about a four-second rate. As the battery charge increases, the *on* time of the green LED decreases and the *off* time increases. A fully charged battery may show green pulses as short as a half-second and the time between pulses may be 60 seconds or more.

T1, D1, C1 and C2 form a stan-

dard full-wave-bridge power supply providing an unregulated 20 V dc at 500 mA. U3, an LM78L12 three-terminal regulator, provides a regulated 12-V source for the control circuits.

Note that the mounting tab on U1 is not at ground potential. U1 should be mounted to a heat sink with suitable electrically insulated but thermally conductive mounting hardware to avoid short circuits. Suitable mounting hardware is included with the PC board (see Note 4).

Other Bulk-Charge Rates

The maximum bulk-charge rate is set by the value of R3 in the series regulator circuit. The formula used to determine the value of this resistor is $R_{ohms} = 1200 / I_{mA}$. T1 must be capable of supplying the bulk charge current and U1 must be rated to handle this current. The LM317T used here is rated for a maximum current of 1.5 A *provided* it has a heat sink sufficiently large enough to dissipate the generated heat. If you increase the bulk-charge rate, you'll definitely need to increase the size of the on-board heat sink. Mounting U1 directly to the housing (be sure to use an insulator) may be a good option.

Transformer Substitution

I selected T1 because of its small size and PC-board mounting. You can substitute any transformer rated at 15 or 16 V ac (RMS) at 500 mA or more. You may find common frame transformers to be more readily available. You can mount such a transformer to an enclosure wall and route the transformer leads to the appropriate PC-board holes.

Construction

There is nothing critical about building this charger. You can assemble it on a prototyping board, but a PC board and heat sink are available.[4] The specially ordered heat sink supplied with the PC board is $1/4$-inch higher than the one identified in the parts list and results in slightly cooler operation of U1. The remaining parts

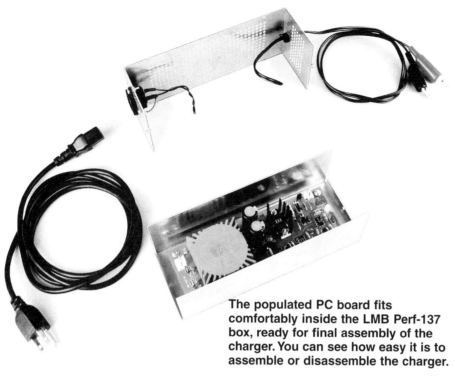

The populated PC board fits comfortably inside the LMB Perf-137 box, ready for final assembly of the charger. You can see how easy it is to assemble or disassemble the charger.

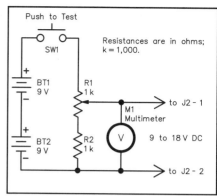

Figure 2—Test voltage source for the battery charger. (The component designation for the push-button switch differs from QST style.)

are available from Digi-Key.

Be sure to space R1 and R3 away from the board by $^1/_4$ inch or so to provide proper cooling. R13 can be a single-turn or a multiturn pot. You'll probably find a multiturn pot makes it easier to set the cutoff voltage to exactly 14.5 V.

R13 Adjustment

To check for proper operation and to set the trip point to 14.5 V dc, we need a test-voltage source variable from 12 to 15 V dc. A convenient means of obtaining this test voltage is to connect two 9-V transistor-radio batteries in series to supply 18 V as shown in Figure 2. Connect a 1-kΩ resistor (R2) in series with a 1-kΩ potentiometer (R1) and connect this series load across the series batteries with the fixed-value resistor to the negative lead. The voltage at the pot arm should now be adjustable from 9 to 18 V. During the following procedure, be sure to adjust the voltage *with the test supply connected to the charger at J2* because the charger loads the test-voltage supply and causes the voltage to drop

a little when it's connected.

Remove the jumper at J1 and apply ac line voltage to the unit at J3. Turn R13 fully counterclockwise. D3 should glow green. Connect the test voltage to J2 and adjust R1 of Figure 2 for an output of 14.5 V. Slowly adjust R13 clockwise until D3 glows red. To test the circuit, wait at least four seconds, then gradually reduce the test voltage until D3 turns green. At that point, the test voltage should be approximately 13.8 V. Slowly increase the test voltage again until D3 turns red. The test voltage should now read 14.5 V. If it is not exactly 14.5 V, make a minor adjustment to R13 and try again. The aim of this adjustment is to have D3 glow red just as the test voltage reaches 14.5 V.

To test the timer functioning, remove the test voltage from J2 and set it for about 15 V. Momentarily apply the test voltage to J2. D3 should turn red for approximately four seconds, then turn green. The regulator is now calibrated and ready for operation. Remove the test voltage and ac power and install the jumper at J1.

A Suitable Enclosure

I used an 8×3×2.75-inch LMB Perf-137 box (Digi-Key L171-ND) to house the charger. An alternative enclosure is the Bud CU482A Convertabox, which measures 8×4×2 inches (available from Mouser). If you use the Convertabox, be sure to add some ventilation holes directly above the board-mounted heat sink. The LMB Perf box comes with a ventilated cover. If you are inclined to do some metal work, you could build your own enclosure using aluminum angle stock and sheet and probably reduce the size to perhaps 8×3×2 inches. If you use a PC-board-mounted power transformer, watch out for potential shorts between the transformer pins (especially the 120-V ac-line pins) and the case. If you use a metal enclosure, connect the safety ground (green) wire of the ac-line cord directly to the case.

Operation

It is very important that this charger be connected *directly* to the SLA battery with no diodes, resistors or other electronics in between the two. The charger works by reading the battery voltage, so any voltage drop across an external series component results in an incorrect reading and improper charging. For example, the Elecraft K2 has internal diodes in the power-input circuit, so it's necessary to add a charging jack to the transceiver that provides a direct connection to the battery. Now I can leave my K2 connected to the charger at all times and be assured that its internal battery is fully charged and ready to go at a moment's notice.

Notes

[1]Larry Wolfgang, WR1B, "Elecraft K2 HF Transceiver Kit," Product Review, *QST*, Mar 2000, pp 69-74.
[2]Although this charger was designed specifically for use with the Power-Sonic PS-1229A SLA battery used in the Elecraft K2 transceiver, its design concepts have wide ranging applications for battery operated QRP rigs of all types.
[3]Although it's labeled a 12-V battery, the terminal voltage is nominally 13.8 V with no load.
[4]A PC Board (double sided, plated through holes, solder masked and silk screened) and heat sink are available from Intelligent Software Solutions, PO Box 522, Garrisonville, VA 22463-0522. Price: $18 plus $1.50 shipping in the US and Canada.

Bob Lewis, AA4PB, became interested in Amateur Radio during junior high school in the late '50s. With the encouragement of his cousin, Al Krugler, K8DDX, Bob obtained his Technician license (K8KNI) and spent most of his time on 6-meter AM in the Detroit, Michigan area. His early interest in Amateur Radio resulted in a career in electronics, first as a radio mechanic in the air-transport industry, followed by ten years in the Navy as an aviation electronics technician. While in the Navy, Bob found 6-meter activity to be a bit sparse in the middle of the Atlantic Ocean, so he upgraded to General, then Advanced and finally, Extra class. He enjoys QRP, PSK31 and home-brewing. Bob is retired from Civil Service, currently working part-time for an electronics consulting firm. You can contact him at Box 522, Garrisonville, VA 22463; **rlewis@staffnet.com**.

By Mike Bryce, WB8VGE

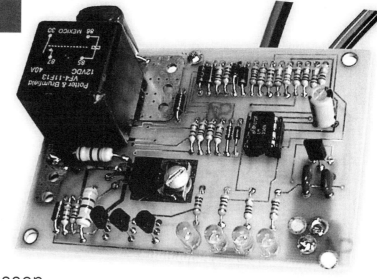

The Protector

Have you ever connected power backward, had a power supply regulator fail or intermittently lose the dc input line to a rig?
If so, you learned an expensive lesson.

Imagine you're working Field Day and unbeknownst to you, a wire comes loose from the battery that is being charged by a solar panel. Instead of the radio seeing 12 V, it's now looking at an unclamped 22 V. Can you see the smoke pouring out of the rig?

Some ac line-operated power supplies have a crowbar circuit to protect the radio from overvoltage. A common crowbar scheme uses an SCR to short the 12 V dc line to ground, blowing the output fuse in the process. The fuse pops, the radio is safe and everyone is happy. Older ac line operated supplies lack over-voltage protection. Do you really feel like digging around inside that supply to retrofit a voltage crowbar circuit?

Have you ever tried to connect your radio to a battery and forgotten to turn the radio off first? There are sparks, the radio flashes on and off and the microprocessor goes for a wild ride until the supply leads are tightened to the battery posts. Let's see… what steps are required to reset the processor and what other damage has been done to the radio?

A solution to these problems was clearly needed. I wanted to protect the radio from an overvoltage condition and from reverse polarity, as well. While I was at it, a delay start to protect the equipment from turn-on transients would also be an excellent idea. After a few weeks of tinkering, the result is a project I call The Protector.

Description

The Protector is a small circuit that can be installed between any 12 V radio and its 12 V dc power source. The power source can be anything from an ac line-operated supply to a battery. It will protect the radio against overvoltage and from reverse polarity. It will also provide a delayed start, just in case you forgot to turn the radio off before you connected it to the power source.

The Protector will protect any load against reverse polarity and over-voltage, and will do so without blowing fuses or popping circuit breakers. In either case, once the problem is fixed, The Protector will once again allow power to be sent to the load. There are no reset buttons to push, no circuit breakers to reset and no fuses to replace. By design, the operation is totally automatic and requires no user input—it is completely transparent. Best of all, The Protector is easy to build and it requires no adjustments or setup.

You should be able to assemble the circuit in about an hour.

All parts mount on a double-sided PC board the size of a playing card. The circuit can be built on a piece of perforated board and hand-wired. To make building easy, however, you can purchase a complete kit of parts, including the PC board.[1]

The Circuit

Relay

The most difficult problem to overcome was switching the relatively large dc required by modern equipment. I tried all sorts of solid-state bipolar and FET switching devices. While they seemed to work, they were overly complex and expensive. I fell back to something simple, easy and inexpensive—a relay. The relay used is a high current device that can easily handle up to 40 A of contact current. With a 12 V dc coil, it's easy to drive with a power MOSFET.

When power of the correct voltage and polarity is applied to The Protector, after the delay start has timed out, transistor Q2 receives base current from diode D1. This causes Q2 to conduct and pull the base of Q3 low, which turns Q3 on. Q3 then sends 12 V to the gate of the power MOSFET, Q4. With the power MOSFET on, the relay coil is energized, the contacts close and power is sent to the radio.

Although the relay has a 12 V dc coil, the relay coil runs hot at 13.8 V. To keep it from overheating and to reduce the overall current draw, two 22 Ω, 1 W resistors are in series with the relay coil.

Overvoltage Shut Down

If the supply voltage exceeds 15 V, Zener diode D2 conducts. This applies base voltage to Q1; it conducts and deprives Q2 of base voltage. When Q2 is off, the base of Q3 goes high and no longer supplies voltage to the gate of the power MOSFET, Q4. The power relay opens, removing power from the radio. The 15 kΩ resistor between the gate of Q4 and ground ensures a speedy dropout of relay K1.

Murphy's law says that any overvoltage condition will occur when the radio is transmitting and drawing maximum current. Arc suppression when the relay contacts open is handled by C4, a 0.47 µF capacitor.

[1]Notes appear on the last page of this article.

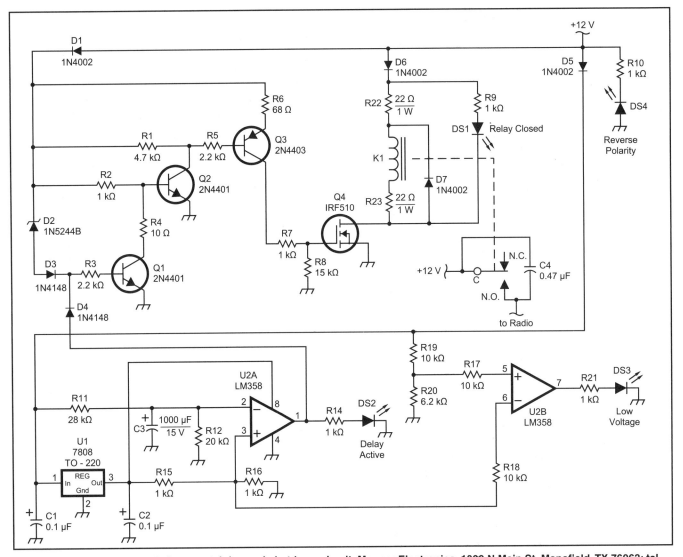

Figure 1—The Protector—a dc input watchdog and shutdown circuit. Mouser Electronics, 1000 N Main St, Mansfield, TX 76063; tel 800-346-6873; www.mouser.com. Digi-Key Corporation, 701 Brooks Ave S, Thief River Falls, MN 56701; tel 800-344-4539; www.digi-key.com.

C1, C2—Capacitor, 0.1 µF, ceramic disk, Digi-Key P4525-ND.
C3—Capacitor, electrolytic, 1000 µF, Mouser 140-XRL16V1000.
C4—Capacitor, 0.47 µF, metal film, Mouser 581-BT151K474.
DS1-DS4—LED, Mouser 351-5502.
D1, D5-D7—Diode, 1N4002, Mouser 583-1N4002.
D2—Diode, 1N5244B, Mouser 583-1N5244B.
D3, D4—Diode, 1N4148, Mouser 583-1N4148.

K1—Relay, SPDT, 12 V dc, 40 A, Digi-Key PB229-ND.
Q1, Q2—Transistor, NPN, 2N4401, Mouser 512-2N4401.
Q3—Transistor, PNP, 2N4403, Mouser 512-2N4403.
Q4—FET, power, N-channel, IRF510, Mouser 570-IRF510.
R1—Resistor, 4.7 kΩ, ¼ W.
R2, R7, R9, R10, R14-R16, R21—Resistor, 1 kΩ, ¼ W.
R3, R5—Resistor, 2.2 kΩ, ¼ W.

R4—Resistor, 10 Ω, ¼ W.
R6—Resistor, 68 Ω, ¼ W.
R8, R12—Resistor, 15 kΩ, ¼ W.
R11—Resistor, 27 kΩ, ¼ W.
R13, R17-R19—Resistor, 10 kΩ, ¼ W.
R20—Resistor, 6.2 kΩ, ¼ W.
R22, R23—Resistor, 22 Ω, 1 W.
U1—Regulator, 8 V, 7808, Mouser 513-NJM78-8FA.
U2—Dual op amp, LM358, Mouser 511-LM385AN.

Start-up Delay Circuit

When power is first applied to The Protector, diode D5 allows capacitor C3 to charge through R11. The voltage at the junction of R11 and C3 is monitored by U2A, an LM358 operational amplifier. As long as the voltage at R11 and C3 is lower than the reference voltage, U2A's output is high and applied to the base of Q1 via diode D4, keeping Q1 on. Q1 thus turns off Q2, deactivating the power relay.

When the voltage at R11 and C3 exceeds the reference voltage (established by R15 and R16), U2A's output goes low. With base voltage removed from Q1, the power relay turns on. Changing the value of C3 and/or R11 will determine the start up delay.[2] Resistor R12 provides a discharge path for C3. This

ensures a new time delay whenever power is removed and then reapplied.

Low-voltage Warning

That second op amp in the LM358 would have been idle, doing nothing, so I used the leftover amplifier as a low-voltage detector.

Resistors R19 and R20 set the reference voltage for the low voltage warning. When the input voltage drops between 11 and 10 V, the output of U2B goes high and illuminates DS3—the only function of this circuit. If you wish, you can leave the associated parts out of the circuit. Since the majority of new microprocessor controlled radios shut down when the supply voltage falls under 12 V dc, you may want to tinker with the values of R19 and R20. The relay will drop out when the supply voltage is below 9 V dc—a cheap and dirty low voltage disconnect.

Reverse Polarity Protection, Status LEDs and Power Supply

In case of reverse polarity, diodes D1, D5 and D6 keep everything off, including the delay circuits. If the supply is connected backward, DS4, the reverse polarity LED, will illuminate. DS1 illuminates when the relay coil is energized to let the user know the power relay is activated.

When power is first applied and the delay circuit is running, DS2 is illuminated. During a low voltage condition, DS3 will light.

The LM358 requires a stable supply voltage. U1, a 7808 regulator, does the job. The input and output leads are bypassed for stability.

Building The Protector

There's nothing special about the circuit layout. You can use point-to-point wiring or dead bug construction if you want. As The Protector is doing a critical job, however, it is best assembled using a printed circuit board. I suggest you forego the use of an IC socket for U2...a socket will eventually come back and haunt you with intermittencies, especially in a field environment.

The power MOSFET, Q4, is sensitive to static discharges. A wrist strap is a good idea. Don't install Q4 until the gate resistors are installed. Once Q4 is installed and on the board, it is very robust and should last forever.

Q4 should be installed so it lies flat against the board. This prevents it from moving around and possibly breaking its leads. Q4 does not require any heat sinking. A 4-40 screw, lock washer and nut hold it to the board.

The LEDs are mounted so they are flush with the board. This prevents LED movement and lead breakage. If you mount The Protector into a power supply with limited space, then I would suggest mounting it on edge. If you do, bend the leads of the LEDs 90 degrees and mount them so they are visible.

Part Substitutions

Although the schematic shows U1 as a TO-220 case, you can use a low-current 78LO8 regulator. Other voltage regulators with outputs from 5 V to 10 V can be used for U1. You would need to adjust the resistor values used in the low voltage warning and reference voltage dividers accordingly, however.

Any diode in the 1N4000 series will work in The Protector. Most common PNP and NPN transistors will work, as well.

I used green LEDs for DS1 through DS4 because I have many of them. You could use LEDs of different colors for the different functions of The Protector.

The power relay is soldered directly to the PC board. A plug-in version of the same relay is available.[3] You would need a socket, however. If you do use it, a center pin must be cut so that the socket will fit the board.

Checkout

An adjustable power supply capable of producing up to 20 V dc will be required for testing. Set the supply between 12 and 14 V. Connect the positive and negative leads of The

Protector to the power supply. Just in case there is a wiring error, do not connect a radio yet.

Turn on the supply. The delay LED (DS2) should light and relay K1 should remain open. After about 10 seconds, the delay LED should go dark and you should hear K1 click on. Anytime K1 is on, the power on LED (DS1) will also light.

Increase the power supply voltage. At approximately 15 to 16 V, K1 should drop out. Lowering the voltage back down to 13 V should cause K1 to click back on and the power on LED and DS1 to illuminate.

Power down the supply and reconnect the leads backward. Apply power once more. Nothing should happen, except that the reverse polarity LED, DS4, should illuminate.

Reconnect the leads with proper polarity and reduce the supply voltage to 10 V. Between 11 and 10 V, the low voltage LED (DS3) should light. That voltage is dependent on the resistors in the two voltage dividers.

Troubleshooting

There's very little that can go wrong. If you can't get The Protector to work, check first for poor soldering and solder bridges. If the soldering is fine, look for parts installed in the wrong location. Be especially mindful of diodes D1, D5 and D6. If any one of the three is in backward, the circuit won't function.

Problems With the Delay Start

When power is first applied, monitor the C3 and R11 junction and see if the voltage increases. If it does, check pin 3 of U2—it should be 4 V. Diode D4 may be defective or installed backward.

Selecting the Overvoltage Trip Point

The voltage of Zener diode D2 determines where The Protector will trip. The voltage drop across diodes D1 and D3 adds an extra 1.4 V to the trip point. That's why a 14 V Zener is used for D2. With D1's voltage drop taken into account, the trip voltage will be about 15 V within a tolerance of 10 to 20 percent. If you want to increase the trip voltage, increase the Zener voltage. Conversely, if you want to lower the trip voltage, select a lower voltage Zener. [B-series Zeners have a 5% tolerance.—*Ed.*]

RadioShack sells a package of two 15 V Zener diodes. When the voltage drop of D1 is taken into account and if you use these,

Figure 2—The Protector shoehorned into a Heathkit HP-1144 power supply.

the relay will drop out at about 16.5 V dc.

Using a 12 V Zener diode from RadioShack, the drop-out point is 13.8 V. I have one of the circuits installed in a Heathkit HWA-202 power supply. The power supply voltage is 13.5 V dc. The Protector has never falsely tripped even though its trip point is only 0.3 V above the supply voltage.

Low Power Operation

Running portable usually means operating from a battery. If you plan on using The Protector in the field with a battery, you might want to swap out the relay used for K1. The relay specified consumes about 60 mA in the circuit as shown. LED DS1 consumes another 20 mA, or so. The low power fix for the LED is simple: simply remove it or R9 from the circuit.

The relay is a bit more difficult. There is no low coil current direct replacement for K1. Instead, use a smaller relay having a lower coil current and contact rating. Most low-power rigs won't require much more than 5 A of dc input, so pick a relay with a contact rating to suit your transceiver's needs. The relay can be mounted to the board dead bug fashion using epoxy.

Life with The Protector

They say nothing is idiot proof, and The Protector is no exception. That's why there are no terminal strips to attach wires to. All the wiring is directly to and from The Protector and the wires should be soldered to the board. The use of appropriately gauged black and red power cord is recommended.

If you plan on using The Protector inside an older ac power supply, route the plus lead through The Protector and out to the radio.[4] The negative leads do not have to run through the board. However, you must supply the circuit with a ground connection. A short length of hookup wire is all you need.

Connect it between the negative lead (the chassis is usually tied to ground and thus to the negative lead) and to one of the ground points on the circuit board. Figure 2 shows an example of The Protector installed in an ac to dc power supply.

If you need to isolate The Protector from the power supply chassis, use nylon hardware to mount the board.

For use in the field, I suggest you purchase or make a dedicated power cable for the radio. Install The Protector between the radio and power source.

Whether your power supply malfunctions—or *you* do by making a serious wiring mistake—The Protector will always keep an eye out for you.

Notes

[1]A complete kit of parts including a PC board is $30 + $4 shipping from SunLight Energy Systems, 955 Manchester Ave SW, North Lawrence, OH 44666; tel 330-832-3114, 888-476-5279 (orders only); **www.seslogic.com**.

[2]The 10-second delay seemed too long when used with an ac supply. You can drop the delay by a factor of 10 by swapping C3 (1000 µF) with a 100 µF capacitor. If you want no delay, remove this capacitor from the circuit.

[3]Digi-Key #PB232-ND is the socket. PB316-ND is a wiring harness so you can locate the relay off the PC board.

[4]If you install The Protector inside a Heathkit HP-1144 supply or any supply with remote voltage sense, disable the delay start. Since the remote voltage sense wire from the regulator is connected to the radio power jack and the delay circuit is active, the relay will stay open until the delay times out. Because the remote sense line is not connected to the output of the supply, the uncontrolled regulator drives the voltage up to 24 V. The Protector sees the +24 V as high and does not close after the time delay. Disabling the delay start insures that the relay pulls in before the regulator senses that the remote sense line is not connected.

You may contact the author at 955 Mancester Ave SW, North Lawrence, OH 44666; **prosolar@sssnet.com**. 　QST▪

By Yaniko Palis, VE2NYP

The 12 Volt Pup: A DC Generator You Can Build

Grab a lawn-mower engine and an alternator to build a great 50 A power supply for Field Day or . . .

JAMES ST. LAURENT

Field Day weekend is the best event of the year! I have always loved wilderness camping and almost any other adventure in the wide-open spaces. Coincidentally, my work often involves setting up all kinds of gear at remote locations for short periods of time—sort of a large-scale version of Field Day. Because of these two interests, Field Day has been my favorite event ever since I became a ham, six years ago. Now, thanks to what I have named "The 12 Volt Pup," I can easily generate enough power to operate a 100 W transceiver and plenty of accessories at almost any location I choose.

Generating power at remote locations is burdensome, in both equipment weight and cost. The Pup weighs about 45 pounds without a battery; so one person can handle it fairly easily. All told, expect to tote anywhere between 70 and 100 pounds, including batteries, fuel, oil and cables. If needed, you can easily disassemble the Pup into assem-

blies weighing less than 20 pounds each for backpacking.

The 12 V Pup combines a standard 3.5 horsepower lawn mower engine with an automotive alternator. These two components mount face downward onto two parallel, heavy duty, **L**-shaped steel rails, as shown in Figure 1. Spacers between the components and the rails precisely locate the pulleys and belt within the two steel rails. (See Figure 2.) Thus, the unit can rest on any

appropriate flat surface. The engine takes a pulley for standard **V** belts, which makes it compatible with the alternator. Add a car battery and presto! You're in business. This design is amazingly simple.

An emergency version of this device could be jury rigged in an hour and a half. All you really need is a pulley for the engine, the right

¹Notes appear at end of article.

Figure 1—A bottom view without protective shields indicates the simplicity of the basic design. An engine bracket is visible at the left end of the lower (front) rail. The slot to mount the alternator (small pulley with fan) is in the upper (rear) rail. The Pup has wooden handles at each end for carrying.

Figure 2—One of the two small engine brackets is above the pipe-coupling spacer. The engine is at the upper right, the front rail at lower left. See Figure 4 for mechanical details.

Partial Parts List for the 12 V Pup:

Motor (1)—(See Figure 4.) After searching for a used engine, I bought a new, no frills lawn mower (for $99) and kept the engine. Recent models have a safety lever connected to a **KILL** switch on the engine that grounds a *neutralizing wire* to stop the engine. This neutralizing wire connects to the control box ignition switch and protection circuit.

Alternator (1)—(See Figure 3.) The one I used is modified as suggested by the folks at a large alternator-remanufacturing company. They rewound a standard alternator with fewer turns so that its internal regulator activates more often (50 A output). A modified unit should cost $65 to $85. Any standard internally regulated alternator with an internal charge controller should be fine, especially for charging automotive batteries. (A used alternator is worth $15 to $30.)

Motor Pulley (1)—Get one sized for standard **V** belts. Its rim diameter should be twice that of the alternator's pulley. This makes the alternator turn twice as fast as the engine. I used a 5^1/$_2$-inch-diameter pulley. It's a big blessing that the engine shaft's dimensions are standard in every way. A common steel pulley fits right onto the engine's ^{7}/$_8$-inch shaft and accepts a standard locking key (^{3}/$_{16}$ inch wide by ^{1}/$_8$ inch deep).

V Belt to fit the pulleys, likely to be somewhere between 27 and 30 inches long; see text.

Storage Battery (1)—12 V lead-acid battery, 15 Ah or greater. Automotive or motorcycle batteries work, but a deep-discharge battery that tolerates fast charging is best. (Gel cells require a closely controlled charging regimen.)*

Steel rails (2) of **L**-shaped angle iron. This material is commonly used to support heavy-duty, industrial-grade storage shelves. It is perforated with rows of holes that ease assembly, provide ventilation and reduce its weight. The flanges should be at least 2^1/$_4$ × 1^1/$_2$ inches. The front rail is 18 inches long; the back rail is 14 inches long.

Motor Brackets (2)—Heavy-duty 1×1-inch angle iron. See Figure 4.

Hardware (Nuts, bolts and spacers—all of which may vary):
   (3) Engine-mount bolts, ^{3}/$_8$×16×2^1/$_2$ inches long
   (3) Spacers, ^{3}/$_4$-inch-diameter, 1^1/$_4$-inches-long steel pipe couplings. These spacers place the engine pulley in the same plane as the alternator pulley. Buy longer couplings and/or shorten them as needed to accurately align the two pulleys.
   (2) Alternator mounting bolts to fit your alternator.

*The regimen is described in "A New Chip for Charging Gelled-Electrolyte Batteries," by Warren Dion, N1BBH, in *QST*, Jun 1987, pp 26-29.

---

parts, $125 should get you all the basic ingredients. My Pup took about four days to create. It's great to use the Pup with two or more deep-discharge lead-acid batteries. You can operate with power from one battery while charging the other. Because the Pup will probably charge a battery much faster than you would normally consume the stored energy, the generator may be switched off perhaps half of the time. This conserves fuel and reduces noise pollution.

You could also connect a load directly to the generator—as long as there's a battery connected across the load to stabilize the alternator's output. The engine's little governor works just fine, readily adapting the throttle to changing load conditions. While idling, the Pup provides about 6 A for normal battery charging. A 50% throttle setting produces about 30 A and ensures proper governor performance under varying loads.

Uses for the Pup go far beyond powering radios: I have inspired a friend to make one for his remote mountain cabin; it's a reliable supplement to his solar panels. A Pup

size belt and two angle iron rails fitted with simple little mounts. Of course, you must also be willing to critically amputate your car and lawn mower! I decided to build a dedicated unit instead; it sports a control box and it cost me only $250 for all new parts. If you can scrounge up used

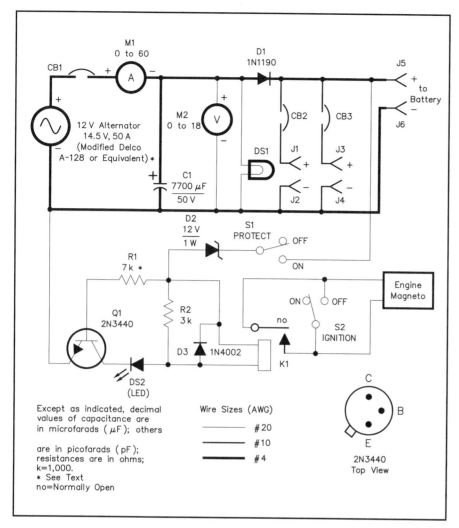

Figure 3—Control box schematic. Equivalent parts may be substituted for those shown. Many of the parts that carry large currents are not available from typical electronic-part suppliers. You'll have better luck at auto-part stores and local electrical-supply shops.

DS1—Automobile panel lamp, 12 V, 6 W, with socket and switch

C1—7700 μF, 50 V aluminum electrolytic

CB1—50 A automotive automatic-reset circuit breaker (from author's junk box; see Note 1)

CB2, CB3—30 A dc circuit breaker switches (65 V dc, 37.5 A trip, No. UPL1-1 from Philips Technologies Airpax Protector Group, 807 Woods Rd, Box 520, Cambridge, MD 21613-0520; tel 410-228-1500, fax 410-228-3456)

J1-J4—30 A terminals or connectors (builder's choice)

J5, J6—50 A terminals or connectors (builder's choice, look at your car's alternator connectors for ideas)

can charge vehicle batteries in the field. The Pup is also an excellent auxiliary power unit for an RV or at the race track, for deluxe golf carts and—my most ingenious use thus far—to charge batteries for electric trolling motors. "Ahoy, mateys! Let's visit a maritime mobile, haar!" I'm sure you'll find a use for a VE2NYP 12 Volt Pup.

Voltage Regulation

Cars do not run on 12 V, and regulated alternators are inherently unstable. Without some additional regulation, even a so-called "internally regulated" alternator will likely put out ugly inductive spikes at a dangerous 20 V, or more. Without other provisions to condition the output, a sizable lead-acid battery is essential; it should stabilize the output to a ripple-free 14.5 V.

The Control Circuit

The control box that I built is very simple. (See Figure 3.) The entire circuit is protected by an internal, auto-

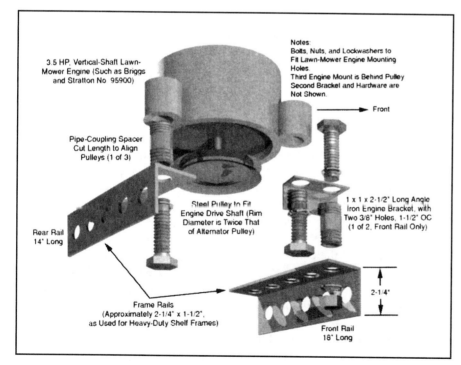

Figure 4—A pictorial of the engine mounting details.

motive, 50 A automatic-reset circuit breaker.[1] The two auxiliary outputs are each protected by 30 A breaker switches. Even with these breakers, this system is as hazardous as that of a car: Shorting the battery, alternator or internal wiring will cause a *big* explosive spark. (They might hear it in Calcutta, but we no longer send code like this!) Carefully avoid electrical shorts at all times—*especially* when handling the battery cables.

To filter the alternator's output, I installed a 7700 mF electrolytic capacitor across it. The capacitor absorbs the output spikes, leaving a rounded reverse-ramp wave as ripple at 0.40 V (a barely tolerable 3.5%). A 6-W panel lamp acts as a minimum load that protects the battery against overcharging. D1 is a high-current blocking diode. It prevents battery discharge through the lamp and reduces the voltage at the battery to about 13.8 V. I also built a very simple protection circuit that stops the engine should the output exceed 15.5 V (16.0 V peak ac).

During its brief life as a prototype, I have already received many good suggestions on how to improve my control box. For instance, one could stay on an automotive theme and use a ballast resistor, solenoid and an ignition relay to disconnect the battery. You could use a heavy-duty headlight switch with an internal circuit breaker for the power switch.[2] All this is to say, the control-box circuit that I show here is only one of many possibilities— you're welcome to improvise!

Finally, I recommend that you study the unit's output with an oscilloscope to be certain that your valuable equipment won't be damaged if the battery is disconnected while you are running the Pup. Also, some 12 V-only devices might be damaged by the 13.8 V dc that this device normally generates. [Most equipment built for automotive use is rated to +15 V.—*Ed.*]

Potential Hazards

There are mechanical dangers from the belt, pulleys and other moving parts. It is *your* responsibility to install adequate mechanical shields to prevent bodily harm. The photos show some metal shields and a plywood base that enclose the moving parts. Cut and fit similar shields to your Pup when the main construction is done. Keep fingers, hair, clothes, etc, completely away from all moving parts.

As with all combustion-powered generators, stray sparks may ignite the fuel. Stop the engine to refuel, and don't start it again until any spills have evaporated. Keep all cables, connectors, switches and relay contacts away from the fuel tank, and use this device only in well-ventilated areas. Closely follow *all* of the engine manufacturer's warnings.

Construction

The exact configuration of your Pup will depend on the actual engine and alternator pair that you acquire. That selection will determine the control-box size limitations. (I temporarily assembled the major parts several times to determine the final arrangement.) These notes may ease your construction. A socket set, wrenches and nut drivers turn this process into a breeze. So tune in your favorite listening frequency and enjoy the pleasures of being an insatiable tinkerer.

As you build, take measures against hazards: Prevent access to moving parts; tighten and seal connections against vibration; allow engine and alternator heat to escape; provide ventilation for cables and contacts carrying high currents; plan for exposure to the weather. Use plenty of grommets, wire ties, heat-shrink tubing, hot glue and strain reliefs to render all the connections Murphy proof.

Soldered connections may melt at the current levels found in this project. I crimped—and then soldered—heavy-duty lugs onto all the cable ends. For high-current connections, I bolted the lugs to the various components and jacks. Almost any circuit that shorts in the control box will likely melt. Finally, keep in mind that your Pup will probably operate in wet environments, so paint and seal its controls and connections against rain (and fuel vapors!).

Mechanical Assembly: Be an Iron Worker in your own Home

In the following assembly notes, I call the side of my engine with the fuel tank and carburetor on it the "front." The spark plug therefore sticks out of the right side and the crankcase is on the left. The alternator is to the left of the engine, beside the crankcase. This places the alternator on the cooler side of the engine (away from the cylinder). The control box is mounted atop the alternator.

Most lawn-mower engines seem to have the same three reinforced mounting holes on their base. (See Figure 4.) Two of the three holes line up with the front, so the long rail goes there. The third hole is at the "rear" and the shorter rail bolts to it. The engine mounts—via two angle-iron brackets, bolts and spacers—to the narrow flange of each main rail; the wide flanges become the vertical sides of the Pup's base. (Refer to Figure 2.)

Before you attach the rails, assemble the two engine brackets to the two front mounting holes on the engine. Position them to point away from the engine, toward the front. These brackets create plenty of elbow room for the engine's new pulley and permit easy access to the oil drain plug. They can swivel slightly, to easily mate with existing holes on the front rail.

Temporarily install the small mounting brackets to the engine, and measure the spacer length (Figures 2 and 4) required to perfectly align the two pulleys. Attach the two main rails so that they extend toward the left as far as possible. It is advantageous that the back rail has only one engine mount because the rail can pivot to accommodate alternators of any diameter.

My alternator did not require spacers because its two mounting holes are flush with the pulley side of its casting. The alternator's cooling fan blades scraped the edge of the rails so I trimmed the blade corners slightly. The threaded mounting hole of the alternator sits on the back rail and mounts to a slot you will cut out of the back rail later. The plain hole on the alternator casting pivots on the front rail, where it's attached. Check all clearances, and ensure once more that the two pulleys are in *perfect* alignment. Verify that the rails and spacers support the pulleys above the ground.

Now measure the arc that the alternator must swing along the back rail to accept a standard-length belt. A slot about 2 inches long allows for a $1^1/_2$-inch variation for belt size, eg, to accept *either* 28 *or* 29-inch belts. (I finished the unit before buying a belt—keep Murphy at bay, I say.) You can plan for standard-length belts during construction using the following formula:

$$BL = 1.57(D + d) + \sqrt{(D+d)^2 + 4C^2}$$
(Eq 1)

where

    BL = Belt length (make all measurements in inches)

    D = Diameter of large pulley

    d = Diameter of small pulley

    c = Distance between pulley centers

To use all available space, I installed the control box on simple rubber-damped mounts that I improvised. They poise the box about $1^1/_2$ inches above the alternator. This allows for air flow and protects the alternator from the rain. Once you have measured all the large internal components and cabling and have established the placement of the control box, pick a suitable cabinet and mark it for machining.

To finish, I picked a spot for a heavy-duty ground lug on the front rail. Thereafter, a few inches will remain open at the left end of the two rails. You can secure a small piece

Figure 5—A rear view clearly shows the largest mechanical shield in place and the carry handle—made from L brackets—that protects the spark plug from damage.

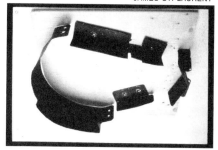

Figure 6—The protective shields, arranged on the plywood base to approximate their mounting positions.

of wood to them, to grasp when lifting the Pup by its left side.

Time to bend, bang, drill, flatten (bang some more), file and sand everything into its final shape. Polish all mechanical grounding points including the engine mounts. Cut the slot out of the back rail with a jigsaw. File off all sharp edges. When the relentless din of power tools, files, twisted blades and flying metal bits finally subsides, you will emerge victorious—and ready for subassembly and painting. Spray paint the mounts, rails and the control box with high-temperature engine enamel.

The protection circuit is built on a piece of perf board. When the output voltage exceeds 15.5 V, a heavy-duty, 5 V PC-board relay grounds the engine's magneto neutralizing wire to stop the engine.

The correct value for R2 depends on the relay's characteristics, so it must be set for each particular relay. To do so, install 10 kΩ pots in place of R1 and R2. Set both pots for maximum resistance. Connect an 18 V variable-voltage power supply across the circuit. (Connect the positive lead to D2's cathode and the negative lead to Q1's emitter.) Set the supply to your desired trigger voltage, and switch on the power. Adjust the R1 pot until the LED just lights. Then adjust the R2 pot until the relay just closes. The two adjustments may interact. Make a final adjustment of R1 when the Pup is complete with the

control box installed and the battery disconnected. Finally, remove the pots, measure their values and replace them with combinations of fixed resistors.

Once my basic unit was tested, I added a pair of modified L brackets with a wood handle to the engine's right side. Together they span over the spark plug to protect it from being broken. (Do *not* loosen the cylinder head bolts to mount this!)

The protective mechanical shields that work well on my particular version are four custom-shaped pieces (cut from 22-gauge sheet metal stock, 7×24 inches). Machine screws hold them to the rails. (See Figures 5 and 6.) Attach the entire unit to a solid base (I used plywood) that blocks any access to the underside of the Pup. Editor Robert Schetgen, KU7G, suggests a lightweight hand cart as a base. Again, keep *all* the moving parts *completely* shielded!

You will love the 12 Volt Pup! It charges big batteries in a couple of hours. A gallon of gas lasts about four hours with a constant 20 A output. It usually loafs at low speed once a large battery has taken its initial charge. The gang at the Concordia University, VE2CUA, Field Day site was very interested in the Pup, and they first suggested that I write this article. Many members already have their own models churning in their minds. Richard Allix, VE2ARW, promises a miniature pup, to be born from a weed whacker and a motorcycle alternator. You are certainly welcome to write me with your comments and experiences. Good Health, Good Luck and Great DX from VE2NYP!

Notes

[1]I did not locate a suitable automatic-reset circuit breaker. Manual-reset breakers in that current range (eg, Potter & Brumfield W31X2M1G-50) cost about $20, or more. A large fuse would be less expensive. Automobile manufacturers use a fusible link to protect the alternator output.—Ed.

[2]According to E. P. Rolek, K9SQG's "A Source for High-Current Relays," in Hints and Kinks (p 73) Wal Mart may be a good source for such parts.

Yaniko "Nick" Palis first became interested in radio communications in his early teens. After some 20 years of SWLing, he finally decided to get on the air by becoming VE2NYP in 1990. Nick ran his college's broadcast radio station and designed many high-power laser light shows in their heyday (up to the early 1980s). He was a lighting director for films and television specials and would sometimes design custom electronic special effects for movies. He was a unit and location manager for many years. Yaniko is presently a supervising producer for feature films and television series in international distribution. Amateur Radio has revived all those previous technical interests and put them to good use again! You can reach Nick by mail at PO Box 61 station Place du Parc, Montreal, PQ H2W 2M9, Canada.

By Joel R. Hallas, W1ZR

Emergency Power at W1ZR

Remember the recent Northeast power failure? Operate your station from battery power all the time and be ready for similar emergencies without operator intervention.

The review of the West Mountain Radio RIGrunner in the October 2002 issue of *QST*[1] finally stirred me into action to redo my station's 12 V dc power system. This had been under consideration for some time as a means to allow easy test of mobile and portable equipment. I was also getting tired of all the "wall wart" power supplies that seem to come with more and more accessories (TNC, keyer, RIGblaster, handheld radio charger), all of which seem to tie up at least two outlet receptacles due to their size.

By the time I was finished, I had deployed a complete emergency power system in addition to meeting my original objectives. The ability to operate HF and VHF on emergency power fits nicely into our regional ARES program. And, my recent completion of the excellent Level I ARRL Emergency Communication Course was in the back of my mind as I was doing this. The resulting system was configured in a way that made it a "no break" or "uninterruptible power supply" (UPS), so I don't even know there's a power failure unless I turn on the lights. As I went through the design, interesting issues arose that I wanted to share with others.

The System Concept

The basic concept is simple and can be divided into three pieces, as described below.

1. Provide a proper dc distribution system so that all necessary equipment can be powered by a single 12 V dc source (RIGrunner 4012 with Anderson PowerPole connectors).

2. Provide a 12 V rechargeable (battery) power source suitable for inside use (deep cycle, recombinant lead-acid, as described later).

3. Add a charging system that will keep the battery charged while in use and bring it back from discharge after power

is restored (three-stage, 10 A, automatic).

The system is shown graphically in Figure 1. Some observations are in order. Consider the battery and charger as a replacement for the power supply that you would buy to run your radio from ac mains power. The battery acts a bit like a very large final filter capacitor with the result that a charger need only supply the average current of all the simultaneously operating equipment, while the battery provides for the peaks (much like the change in power supplies required for a 1950s-era 100 W AM transmitter, compared to today's low duty cycle SSB transmitter, if you go back that far!).

In my case, the HF radio needs (key down) a 20 A supply and the VHF radio, 10 A. The various dc powered accessory equipment (keyer, RIGblaster, TNC) perhaps another 1-2 A. If my HF and VHF radios both transmit together (as they do when on VHF packet while operating HF), I would need a 35 A supply to power everything from a single source. With the charger/battery arrangement, however, I need only provide about 5-6 A from the charger to keep the battery supplied during the time I am operating.[2]

A 10 A charger and a 12 V recombinant (described later) deep cycle battery cost more than a 35 A commercial supply, but less than the combined supplies from the two radio manufacturers. Unlike either of those, however, my system continues to operate W1ZR at full performance for about 11 hours without ac power. This is with no switching or reconnecting of cables—in other words,

without operator intervention. Figure 2 shows the operating position at W1ZR.

What About the Battery?

The battery is at the core of this arrangement. For a number of years I had considered doing something like this, but was put off by the byproducts a lead-acid battery emits while under charge (hydrogen gas and a sulphuric acid). This was not a showstopper, but either seemed to require moving the batteries outside the house (with long connections), or the construction of a forced venting system from near the basement shack location to outside. Otherwise, hydrogen gas would fill the corner of my basement and could be detonated by the furnace pilot light, requiring a new ham station at a minimum!

Fortunately, technology came to my rescue. In the past few years, while most hams were watching the development of new solid-state devices, the lead-acid battery makers were quietly having a revolution of their own. There are now several technologies that provide for "recombinant" operation of storage batteries. In a recombinant battery, most of the hydrogen is not released, but recombines with oxygen within the battery to form water. Thus, you not only avoid the threat of explosion, but you never need to add water.

The recombinant technologies are found in batteries labeled AGM (absorbed glass mat), VRLA (valve-regulated lead-acid) or gel cell. These batteries hold the electrolyte against the plates in a way that avoids (but doesn't quite eliminate) the release of

[1]Notes appear at the end of this article

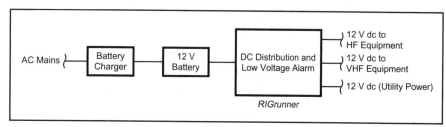

Figure 1—A system diagram of the uninterruptible battery power system at W1ZR.

Figure 2—The operating position at W1ZR.

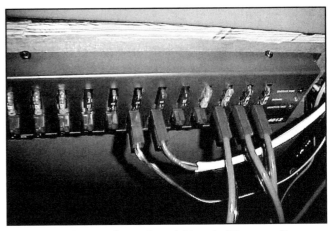

Figure 3—RIGrunner located on right end of the operating position. The battery and charger are on the other side of the wall in the basement utility room.

hydrogen during the charging cycle. A small amount of gas is released, but it is considered sufficiently small that these batteries can be used with normal household ventilation. They are used to power indoor computer UPS systems and motorized wheelchairs, for example. They also do not freeze or spill and will not leak acid.

Fortunately, these batteries are also of "deep cycle" design. A deep cycle battery, unlike the usual auto or marine "starting battery," is designed so that it can be 75% discharged hundreds of times, rather than just a few times, and still be recharged to provide full capacity.[3]

When selecting a battery, watch the description carefully, as not all "sealed" or "no maintenance" batteries are recombinant. Some simply have no ports for water addition and provide a bit more water to start with, but emit all the hydrogen of an open top battery. When the water level finally falls below the top of the plates they start to fail. They are not good for deep cycle use—or for avoiding explosion! Look for "AGM," "VRLA" or "gel cell" if you want to minimize hydrogen emission.

Once you have settled on the battery "family," there is a range of sizes and capacities to consider. The most important parameter for our application is "capacity" in ampere hours (Ah). My battery is rated at 80 Ah. Generally, the higher the Ah capacity, the higher the cost and weight. In an ideal world, 80 Ah would mean I could draw 1 A for 80 hours or 80 A for one hour. Looking at the fine print, one determines that this rating is for a given current. Mine is rated for 80 Ah at 4A, close to our design load.[4]

The RIGrunner has an alarm that can be set to let you know when the input voltage has dropped to 11.5 V. It may be

a good idea to take advantage of that and stop transmitting before everything dies. With my HF transceiver, I received a report that I had keyclicks 100 kHz from my carrier just before the transceiver ceased operation at less than 11 V. I recommend, therefore, that you research the characteristics of your transmitter when it is operated below rated voltage, before you operate at low dc input voltages.

Charger System

The design of most chargers is such that they are current limited at their rated output. Thus, when drawing a load greater than the charger can supply, the excess current will come from the battery, not the charger. Still, the charger output should be fused at its rating (mine is fused at 10 A) to protect the charger from excessive load if something goes awry.

In order to be able to achieve the battery life described above, the charger needs to be able to support multiple phases of recharge, as well as different characteristics for different families of batteries (most gel cells should not be charged above 14.1 V, for example). The following description of a three-phase charger is from the Guest Company, the maker of the charger I selected.[5]

Stage 1: Bulk—When the battery is at 75% capacity or lower the charger pumps high amps at a relatively low voltage.

Stage 2: Absorption—As the battery is charged to 75% capacity, the charger lowers the amps and increases voltage (never exceeding battery's designed voltage maximums) to gradually bring the batteries to full charge.

Stage 3: Maintenance (often called "float")—When batteries are fully charged, the charger drops voltage to a maintenance level and gently maintains

the battery at a full charge.

The alternative, again from Guest literature (no doubt with their product in mind):

Linear Chargers—When the battery is fully charged, units shut off until battery drops to 90% capacity and then turns on to bring it back to full charge. Result— Deep cycles have limited cycles built into them thereby reducing life of the batteries. Other types of batteries are charged at a higher voltage rate, which also reduces life. [There are linear 2 and 3 mode "smart" chargers available, however, that do not use microprocessor control. These use analog comparators to sense voltage and current.—*Ed.*]

For my station, I chose a Guest model 2610 that provides up to 10 A or two independent 5 A outputs to charge two batteries simultaneously. The charge current is applied in the three stages defined above. I have permanently mounted this adjacent to my station. A portable style charger could also be used and moved to the garage, boat or RV, as needed. One downside to a so-called "smart" charger: I note hash around 3.58 MHz, likely due to an internal processor clock. The conducted (radio on dummy load) level was S4, lower than the typical background noise at my location. I reduced it to the internal noise level by using a brute force choke[6] on the dc output. I still note some radiated noise, again on 80/75 meters when the antenna is connected, so shielding all wires and filtering on the ac line side would likely solve the problem. If it proves to be troublesome, a quick fix is to pull ac power off and run solely on dc power. There may be other chargers with similar charging characteristics but without this problem; however, I was not prepared to undertake an exhaustive study of all the available options. Analog

Table 1
Copper Wire System Loss

Wire Gauge (AWG)	Resistance (W /1000 Feet)	20 A Loss for 4 Conductors x 6 Feet = 24 Feet Total (V)	Voltage at Load for 13.8 V Battery Terminal Voltage	Voltage at Load for 12 V Battery Terminal Voltage
8	0.640	0.307	13.5	11.7
10	**1.018**	**0.489**	**13.3**	**11.5**
12	1.619	0.777	13.0	11.2
14	2.575	1.236	12.6	10.8
16	**4.094**	**1.965**	**11.8**	**10.0**
18	6.510	3.125	10.7	8.9
20	10.35	4.968	8.8	7.0

"smart" chargers are available, however, that do not use microprocessor control and hence avoid the clock noise problem. A useful site with several informative application notes is Ibex Manufacturing, Inc.[7] An alternative marine brand with a good reputation is Xantrex[8] (formerly Heart).

Watch that "Copper Loss"!

Everyone "knows" that wire has resistance, however, we are a bit conditioned into thinking that if there is 12 V at one end of a pair of wires, it will also be at the other end. While the difference is slight in the realm of low power stages, we're talking real amperes here with a significant resulting voltage drop. If we have 6 feet of wire between the battery and the RIGrunner and another 6 feet to the radio(s), that's 12 feet of two wires in series or 24 feet of wire resistance to consider. Table 1 illustrates the results for a 20 A load.

I have highlighted the entries for both 10 gauge and 16 gauge wire to make the point. What this says is if your radio draws 20 A and you have it connected via 6 plus 6 feet of 16 gauge wire to a typical 13.8 V dc power supply, it will still see 11.8 V at the equipment and will likely work fine. It may work fine on the battery if the charger is holding the voltage up well above 12, but when the charger is off and the battery is on its way down, it is very unlikely for the radio to work at the 10 V it will see! Move to 10 gauge wire and it is likely to work, but you'd still better check! My deep cycle battery specification says it will deliver 12 V until discharged 50% and then 11.5 V at a 75% discharge level.

Note that in Table 1 that half (for equal lengths) of the voltage drop shown is in the wire between the battery and RIGrunner and half the drop is between RIGrunner and the radio. If we connect a low power accessory to the RIGrunner (for the 10 gauge wire case) the voltage at the RIGrunner will be half the drop shown, or 12 V at the battery, 11.75 V

at the RIGrunner and 11.5 V at the radio. The low powered accessory can use smaller wire from the RIGrunner (11.75 V) to it without a problem. For example, if the accessory draws 0.5 A, the drop in 6 feet (multiplied by 2 wires) of even 20 gauge wire between the RIGrunner and the accessory is only about 0.06 V, so even wire that small should work. This calculation should be made for the current drawn by each of the equipment types.

Please note that if your HF radio draws 20 A and your VHF radio draws 10 A and they will be both transmitting (key down) at the same time, you will have 1.5 times the "drop" shown on the wire from the battery to the RIGrunner, so take that into account, as well. Of course, if you follow the rules and "...use the minimum transmitter power necessary to carry out the desired communications" you may be able to use even less current. Unfortunately, the HF radios that I have checked do not reduce input current as fast as they reduce output power, although that may still be a significant benefit to dc power reduction. This same kind of analysis should also be made for any mobile or shipboard installation, especially if you operate with the engine off.

What can you do to circumvent these pitfalls? Use as large a gauge wire as you can find/afford/terminate/bend. Make it as short as possible. West Mountain Radio[9] provides 10 gauge and 12 gauge wire in various lengths, preterminated, at attractive prices. Marine supply dealers have nice tinned, extra flexible, red/black "duplex" wire in an outer jacket in various sizes. By the way, they also have all the other pieces you need for this project with the exception of the RIGrunner. Auto supply dealers may be another source for parts for the dc system, but I am more familiar with marine dealers in my area. Marine dealers also have "primary" wire in much larger sizes such as 2, 1, 1/0 and 2/0 gauge, for longer runs (to outside battery sheds, for example). If you go to a larger gauge than 10, you

will have to have an intermediate connection block (a "barrier strip," for example) to transition to a size that will fit in a PowerPole to connect to the RIGrunner. Figure 3 is a view of the RIGRunner, installed near the operating table. Figure 4 shows the battery and charger.

Related Issues
Battery Safety

We tend to think of 12 V systems as safe (compared to the 1000 V to 3000 V behind the panels of our linear amplifiers, for example). They certainly are from the point of view of an electrocution hazard, but storage batteries of this sort have significant energy and can do serious thermal damage to people and objects. Our usual 12 V power supply will often "crowbar" to 0 V when shorted. The battery, however, will expend all of its energy in dramatic ways including possibly an explosion! Early in my career I vaporized part of my wedding ring by getting it between a wrench on a battery terminal and a steel floor in a military installation. Since my finger was in the ring, it was quite traumatic. Fortunately both my finger and my marriage survived (40 years to the lovely W1NCY). Please follow the following rules for battery safety.

- Always wear safety glasses when working around storage batteries.
- Do not have open flames near batteries, especially while under charge.
- Never use metallic tools long enough to reach between the battery terminals or connections.
- Install fuses as close to the battery terminals as possible.
- Protect the top of the battery (plastic battery box with lid, for example) so wires or equipment can't fall onto the terminals.
- Wash hands immediately following contact with the battery.
- Use proper size ring terminals on all battery connections. I use crimp-type connectors and solder them after crimping.
- Remove all metallic hand jewelry

(rings, bracelets…see above!).
- Think twice; act once!

Other Applications

You may want to consider other types of loads depending on your environment. A key possibility is lighting. I did consider including dc lighting in my plan, but I've deferred that for the moment. Marine dealers (probably not the cheapest source of dc lights) have various boat cabin lights available. High efficiency focused lights typically draw 2 A, while a 50 W standard (12 V) bulb will draw more than 4 A…almost as much as the radio equipment! This will reduce your operating time by about half. I have some dry battery lanterns I thought I'd use and I may go to some kind of fuel-powered light in the future to save battery energy for the radios.

I also have not yet provided an inverter to generate 120 V ac. This is something that I will consider, but they are known as RFI generators. I don't see them as significant to radio work at my station. In a non-emergency power failure, I can imagine a request to run the refrigerator or furnace from time to time, and that may be a capability worth having if the load is reasonable.

While the TNC is on UPS power, my regular station computer is not. I do have a battery-operated laptop, which can be used for a number of hours once power is off. I have APC UPS systems (model CS500) for both of the main household computers. These do a great job of keeping the machines up for about half an hour and then gracefully shut them down via a serial (USB) connection to the PC. Unfortunately they, too, generate some conducted radio hash, so I opted not to use one for the station computer. There is also a UPS on the house Ethernet hub and DSL firewall router that keeps external Internet communications up while the laptop is operational.

Cost Considerations

This approach is straightforward, but I was surprised how thin my credit card was after buying all this! By the time I was done I could have purchased a new two-band VHF-FM radio for what I spent. If you are not ready to commit to this level of effort or expense, I think you could sneak up on it. Perhaps you already have a charger for other purposes and could find a partially used recombinant battery to use for a few years.

Another thought (thanks to Del Schier, K1UHF) is to use your current radio supply to keep a battery charged (on float). The "float voltage" level of the Douglas battery I have is 13.5 to 13.8 V dc, which

Figure 4—The battery and charger. Note the ferrite choke on the charger leads. The battery is normally covered to enclose its terminals.

is the typical output of an HF radio power supply. As with the charger, you need to confirm that the peak current divides between the battery and supply so that the maximum rating of the supply is not exceeded.

The use of "any old" battery is not recommended for this application—unless you can obtain a surplus recombinant battery from a medical supply house or other source, or you are willing to provide a specially vented battery area. In my opinion, the risk of hydrogen gas explosion is not worth any possible savings.

Notes
[1] S. Ford, WB8IMY, "The RIGrunner," *QST*, Oct 2002, p 59.
[2] I calculated the average current requirements considering the duty cycle of transmit periods. In a typical operating hour, I listen more than I talk and use mostly CW (SSB is similar). My specified receive current for each radio is 1 A (confirmed by measurement). My peak HF current is about 15 A, once tuned. If I actually transmit 10 minutes in an operating hour and with CW used at a 50% duty cycle while transmitting, the result is, for HF: 1 receive × 50/60 + 15 A × 0.5 × 10/60 = 2 A average. For VHF packet, I estimate that there are 30 ten second transmit bursts in an hour (I haven't confirmed this; if you use a DX cluster, you may want to check). During key-down, full power is used. So, for VHF packet: 1 A receive × 57/60 + 10 A × 1.0 × 3/60 = 1.45 A average. (Note that with 2 meter FM voice, the full transmit power is used the entire time the mic button is held down.) The total average current during an operating hour is therefore 2 A + 1.45 A + accessory A = 4.5 to 5.5 A.
[3] For more information about auto and deep cycle batteries, check **www.uuhome.de/ william.darden**, which seems an authoritative source. The battery I purchased was a Douglas DG 12-80 available through

marine dealers. They also have the best specification sheet I have found (**www. douglasbattery.com/gproducts/pdf/ dg12-80.pdf**). This provides all relevant data on their battery's charge-discharge operation, the output voltage as a function of discharge and anticipated life.
[4] For my Douglas DG-12 battery, 80 Ah is specified at a 4 A load or 20 hours of operation at 4 A. At a 7 A load, we can draw current for 10 hours or get 70 Ah. For one hour of discharge, we can only draw 49 A, so the rating drops to 49 Ah. Interpolating, our 5.5 A load should discharge in about 15 hours. When the battery manufacturers talk about discharge, they are specifying the time to discharge to a terminal voltage of 10.5 V. This is the minimum safe terminal voltage for a lead-acid battery. When the per cell voltage reaches 1.7 V (10.2 V for a 12 V battery) permanent chemical and physical damage to the battery can occur. This is probably a lower voltage than we can use, especially considering copper losses (below). If we pick 11.5 V as our minimum usable voltage 75% of the 10.5 V Ah rating should be about right. You can find your minimum voltage using an adjustable supply connected at the battery location and observing when your radios stop being able to transmit ("stop being able to transmit" should be interpreted as "stop being able to transmit *cleanly*") but the receive function will generally continue to a lower battery voltage. The battery will last longer at this rate as well. For 100% discharge, my battery can repeat the charge-discharge cycle almost 200 times (to 60% capacity retention, 100 times to stay at 100%). At 75% discharge, this increases to 300 (200 for 100% retention), 400 at 50% and 1200 cycles for a 30% discharge each cycle. The use of two batteries will permit continued operation during an extended outage, with batteries alternately charged by an automotive or other outside system.
[5] The Guest Company, 95 Research Pky, Meriden, CT 06450; **www.guestco.com/ Chargepro/chargepro.html**.
[6] See any recent *ARRL Handbook* for the description of a "brute force filter." I used a pi-section with 0.01 μF, 50 V (RadioShack 272-131) capacitors on the output of the charger and across the battery. For the series inductance I wound 8 turns (as many turns as I could fit) of 14 gauge "marine duplex" wire (from the charger to the battery) on a CWS (formerly Amidon) FT-240-61 toroid core. A future project will be to make a more permanent version for both dc and ac connections to the charger, mounted right at the charger terminals.
[7] Ibex Manufacturing, Inc, PO Box 294, Francestown, NH 03043; tel 603-547-6209; **www.ibexmfg.com**.
[8] **www.xantrex.com**.
[9] West Mountain Radio, 18 Sheehan Ave, Norwalk, CT 06854; tel 203-853-8080; **www.westmountainradio.com**.

All photos by the author.

Joel Hallas, W1ZR, of Westport, Connecticut, has been an active amateur since 1955. He received a BSEE from the University of Connecticut and an MSEE from Northeastern University, and has been a radar and telecommunications systems engineer for more than 30 years. Joel and his wife Nancy, W1NCY, have two grown children and a golden retriever. He holds WAS, WAC, CP-30, DXCC and DXCC-CW. Joel is the Product Review Editor at QST *and can be reached at* **w1zr@arrl.org**.

West Mountain Radio PWRgate and Computerized Battery Analyzer

Reviewed by Larry Wolfgang, WR1B
Senior Assistant Technical Editor

A friend recently gave me a 12 V, 100 ampere hour (Ah) gelled electrolyte battery that had been pulled from service in a bank of emergency power backup batteries. The commercial communications service my friend works for replaces the entire bank of batteries if one of the batteries drops below about 80% capacity. My main reason for having the battery is to have a power source to run some radio equipment so I can demonstrate ham radio when I go camping with Boy Scout groups.

I needed a convenient way to keep the battery charged and ready to go for weekend campouts. I contemplated various additional pieces of equipment that I might need to safely keep the battery charged and also be able to connect my station equipment to exercise the battery occasionally. I have used a small portable solar panel and a Micro M + charge controller[4] to recharge the battery after summertime Scout operations—not good for winter months.

While looking through a recent issue of *QST*, I noticed an ad for the West Mountain Radio PWRgate PG40. The PWRgate can handle up to 40 A continuously, either from a power supply or battery. That makes it a perfect match for my 40 A switching supply, and will handle the current requirements of my Elecraft K2/100 HF rig, KAT100 antenna tuner and a VHF FM transceiver and other station accessories.

The PWRgate PG40 looks like a small heat sink with three sets of Anderson Powerpole connectors on top. The Powerpole connectors follow what has become the common standard for ARES and other emergency operators. The compact PG40 package is about 4×5×1⅝ inches (H×W×D), and has mounting holes on each end. Connect your 13.8 V dc supply supply to the PS input and your 12 V battery to the BAT input. Then connect your rig or dc distribution panel to the OUT connector and you are ready to operate.

When the ac supply is on, your station power supply powers your equipment. If the ac supply goes off for any reason, the PWRgate instantly routes the battery voltage to the output connector. My radios didn't even blink when I tested the operation by pulling the PS connector out of the PG40. With my K2 display set to read applied voltage and current I could see the supply change instantly from 13.8 to 12.0 and back again when I unplugged and reconnected the ac supply to the PG40. With a PG40 connecting my ac supply and back-up battery, I may not even know the commercial power went off until it gets dark.

What's Inside?

The PWRgate uses two 80 A Schottky diodes wired to isolate the battery and power supply from each other. The input with the higher voltage automatically connects to the output. The Schottky diodes have a forward voltage drop of about 0.4 V that should be considered during your dc wiring design.[5] The PWRgate also uses a diode and resistor to provide up to 1 A of charge current to the battery when the ac supply is operating. When the ac power comes back on, the PG40 will switch back to your power supply and your battery will start to recharge at up to a 1 A rate.

It will take a while to recharge my 100 Ah battery if it is completely discharged. After a relatively brief operating period on the battery, though, the PG40 will top up the battery and keep it fully charged and ready for the next emergency or weekend campout.

How Do You Know That Battery is Taking a Full Charge?

As any battery ages it will tend to lose some of its ability to take a charge and return the energy as useable current. This is where the West Mountain Radio Computerized Battery Analyzer (CBA) becomes valuable. The CBA looks like a large heat sink and top-mounted fan attached to a small plastic tray. It measures about 2¾×3½×3 inches (H×W×D). See Figure 5.

An 8051 microcontroller measures current, temperature and voltage, in three

automatically switched ranges to 10 bit resolution. The microcontroller operates a pulsewidth modulated power MOSFET load for the battery under test. A plug-and-play USB interface to your computer allows software control of the load current, temperature sensed cut-off (with the optional temperature probe) and also allows the computer to collect data about the battery discharge characteristics.

The user interface to the CBA is via *Windows* software that installs from a CD included with the package. You will need a PC running *Windows* 98SE or higher with a Pentium 233 or faster processor, at least an 800×600 display and an available USB port.

The warnings about heat build-up when charging and discharging the batteries point to the value of the optional temperature probe. This optional device plugs into the side of the CBA. When connected, the software will display the temperature. You can set an automatic test cut-off temperature, so if your battery becomes too hot the test will stop. The default temperature is 140°F, but you can change that in the software.

Setting Up a Test

Before you are ready to start a battery test, take a few minutes to read the safety information file on the CD, as well as the CBA owner's manual and the battery information and FAQ files. Then double click on the software icon and a test window will open. Type a descriptive name into the TEST NAME box. Next you will have to select the battery type from a drop-down list that includes most battery types. Enter the battery capacity for the battery you are testing. When you connect the battery to the CBA it will measure the battery voltage and take a guess at the nominal battery voltage and the number of cells

[4]*The ARRL Handbook for Radio Communication*, 2005 edition, pp 17.41-17.44. Available from the ARRL Bookstore for $39.95 plus shipping. Order number 9280. Telephone toll-free in the US 888-277-5289, or 860-594-0355; **www.arrl.org/shop/**.

[5]J. Hallas, "Emergency Power at W1ZR," *QST*, Dec 2003, pp 41-44.

in the pack. The software usually defaults to a test current equal to the capacity of the battery. The software also sets a default test ending voltage; in the case of the 12 V gel cell, the default is 11.4 V.

If you want to use these software defaults, simply click the START button. If you want to find out how long the battery will run *your* equipment, set the test end voltage to the lowest battery voltage at which your radio will still operate (taking into account all voltage drops between the battery and radio) and the current to correspond to the drain of your setup. The graph automatically adjusts the vertical voltage scale and the horizontal capacity scale to suit your test. The CBA continues to draw the specified current until the battery voltage drops to the end voltage.

The CBA can dissipate at least 100 W, higher for lower capacity battery packs. If you try to set a test current that will exceed the ratings, the software will prompt you to reduce the test current to a safe level.

I wanted to know how my 100 Ah gel cell would compare to that rated capacity, so I ran a test on the fully charged battery. See Figure 6. With a 7 A load it has a capacity of about 88 Ah, down to 10.4 V.

I also tested a 17.2 Ah 12 V gel cell that I have used to operate a 2 meter rig for many of my Scout demonstrations over the years. I recharged and tested this battery several times, but it only seems to have a capacity of about 1.2 Ah. I was convinced this battery was better than that, but after several tests I have concluded that it isn't providing as much

operating time as it should. It may be well past time to replace that battery. I also tested another old gel cell that didn't seem to be holding a charge very well. Sure enough, the CBA indicates that battery is beyond any useful life.

To test some known good batteries I decided to set up test fixtures for a few other battery types.

A quick trip to RadioShack produced a single AA cell holder that I clipped to the Powerpole adapter cable to test some AA cells. I have been using some NiCd and NiMH cells for a variety of applications recently, and was curious as to how they tested when compared to an alkaline cell [note that testing the capacity of a non-rechargable battery is a destructive test, but useful on a sample of a battery type—*Ed.*]. Figure 7 shows a graph comparing three different AA cells.

Other Software Features

The CBA software has a number of other very useful features. Obviously, you can save each test graph for later review. You can also overlay one test on top of another. Figure 7 shows three such graphs overlaid on each other. When you overlay a test graph, the software color codes each new graph, and also places a color-coded legend along the right side of the graph. If you have a color printer, you can print the image in color. If you only have a black printer, then the software will add symbols to the lines and legend to distinguish them from each other.

The software also has a *print labels* feature that helps you set up and print labels to apply to your batteries. If you have several similar packs, this is espe-

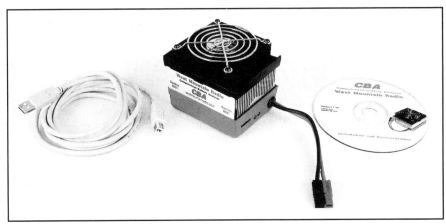

Figure 5—The Computerized Battery Analyzer consists of a heat sink and small fan sitting on top of a plastic tray.

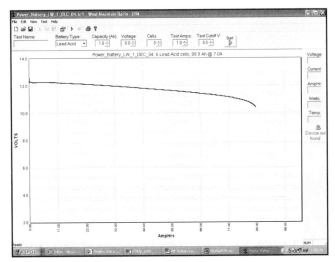

Figure 6—The results of testing my pulled 100 Ah gel cell. I set this test to draw 7 A from the battery. Notice that the test stopped when the battery voltage dropped to 10.4 V. This battery appears to have a capacity of about 88 Ah, rather than the rated 100 Ah.

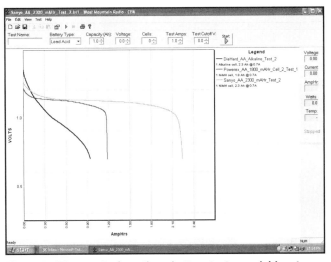

Figure 7—The results from three battery tests overlaid on top of each other. The test compares a Sears DieHard alkaline AA cell with two different NiMH AA cells.

cially useful because it helps you keep track of the individual packs. Rather than always using the same one or two packs, you can easily rotate the several packs and use them uniformly. You can also keep track of which packs are older, so it is easier to decide when to test a pack that may be getting past its useful life. The label template is set up to use Avery 05160 or equivalent label sheets.

How Often Should I Test My Batteries?

That is an important question. If a rechargeable battery is good for a certain number of charge/discharge cycles, then every time you test the battery you can subtract one from the remaining life of your battery. Obviously, over testing is not a good thing. West Mountain Radio suggests that you run a test for a new battery pack [especially good advice for NiCd batteries—*Ed.*]. That will give you a point of comparison as the pack ages. They also recommend that you retest the pack after it has been setting for some time, or after it has been used a great deal. I would probably err on the side of not testing often enough rather than over testing. West Mountain Radio says they do not recommend testing or cycling batteries if they are used regularly and seem to be working well. That makes sense to me.

Testing Accuracy

I used my digital multimeter (DMM) as an ammeter to verify that the actual load current was close to the indicated test current. Within the accuracy of my meter, I measured the same current as the CBA. I also used the DMM to monitor the battery voltage during a couple of tests. Again, within the measurement accuracy of my meter, the CBA software was reporting the same terminal voltage.

The manufacturer reports that by our publication date they will offer a CBA II with higher resolution at low current.

Manufacturer: West Mountain Radio, **www.westmountainradio.com**, 18 Sheehan Ave, Norwalk, CT 06854; tel 203-853-8080. Price: PWRgate PG40, $69.95; Computerized Battery Analyzer, $99.95; Temperature Probe for CBA, $10.95. QST

Index

About the ARRL
The national association for Amateur Radio

The seed for Amateur Radio was planted in the 1890s, when Guglielmo Marconi began his experiments in wireless telegraphy. Soon he was joined by dozens, then hundreds, of others who were enthusiastic about sending and receiving messages through the air—some with a commercial interest, but others solely out of a love for this new communications medium. The United States government began licensing Amateur Radio operators in 1912.

By 1914, there were thousands of Amateur Radio operators—hams—in the United States. Hiram Percy Maxim, a leading Hartford, Connecticut inventor and industrialist, saw the need for an organization to band together this fledgling group of radio experimenters. In May 1914 he founded the American Radio Relay League (ARRL) to meet that need.

Today ARRL, with approximately 150,000 members, is the largest organization of radio amateurs in the United States. The ARRL is a not-for-profit organization that:
- promotes interest in Amateur Radio communications and experimentation
- represents US radio amateurs in legislative matters, and
- maintains fraternalism and a high standard of conduct among Amateur Radio operators.

At ARRL headquarters in the Hartford suburb of Newington, the staff helps serve the needs of members. ARRL is also International Secretariat for the International Amateur Radio Union, which is made up of similar societies in 150 countries around the world.

ARRL publishes the monthly journal QST, as well as newsletters and many publications covering all aspects of Amateur Radio. Its headquarters station, W1AW, transmits bulletins of interest to radio amateurs and Morse code practice sessions. The ARRL also coordinates an extensive field organization, which includes volunteers who provide technical information and other support services for radio amateurs as well as communications for public-service activities. In addition, ARRL represents US amateurs with the Federal Communications Commission and other government agencies in the US and abroad.

Membership in ARRL means much more than receiving QST each month. In addition to the services already described, ARRL offers membership services on a personal level, such as the ARRL Volunteer Examiner Coordinator Program and a QSL bureau.

Full ARRL membership (available only to licensed radio amateurs) gives you a voice in how the affairs of the organization are governed. ARRL policy is set by a Board of Directors (one from each of 15 Divisions). Each year, one-third of the ARRL Board of Directors stands for election by the full members they represent. The day-to-day operation of ARRL HQ is managed by an Executive Vice President and his staff.

No matter what aspect of Amateur Radio attracts you, ARRL membership is relevant and important. There would be no Amateur Radio as we know it today were it not for the ARRL. We would be happy to welcome you as a member! (An Amateur Radio license is not required for Associate Membership.) For more information about ARRL and answers to any questions you may have about Amateur Radio, write or call:

ARRL—The national association for Amateur Radio
225 Main Street
Newington CT 06111-1494
Voice: 860-594-0200
 Fax: 860-594-0259
 E-mail: **hq@arrl.org**
 Internet: **www.arrl.org/**

Prospective new amateurs call (toll-free):
800-32-NEW HAM (800-326-3942)
You can also contact us via e-mail at **newham@arrl.org**
or check out *ARRLWeb* at **http://www.arrl.org/**

FEEDBACK

Please use this form to give us your comments on this book and what you'd like to see in future editions, or e-mail us at **pubsfdbk@arrl.org** (publications feedback). If you use e-mail, please include your name, call, e-mail address and the book title, edition and printing in the body of your message. Also indicate whether or not you are an ARRL member.

Where did you purchase this book?
□ From ARRL directly   □ From an ARRL dealer

Is there a dealer who carries ARRL publications within:
□ 5 miles   □ 15 miles   □ 30 miles   of your location?   □ Not sure.

License class:
□ Novice   □ Technician   □ Technician with code   □ General   □ Advanced   □ Amateur Extra

---

Name _____ ARRL member? □ Yes □ No

_____ Call Sign _____

Daytime Phone ( ) _____ Age _____

Address _____

City, State/Province, ZIP/Postal Code _____

If licensed, how long? _____ e-mail address: _____

Other hobbies _____

_____

Occupation _____

For ARRL use only		EPFRC
Edition	1 2 3 4 5 6 7 8 9 10 11 12	
Printing	1 2 3 4 5 6 7 8 9 10 11 12	

From _____

_____

_____

Please affix postage. Post Office will not deliver without postage.

EDITOR, EMERGENCY POWER FOR RADIO COMMUNICATIONS
ARRL—THE NATIONAL ASSOCIATION FOR AMATEUR RADIO
225 MAIN STREET
NEWINGTON CT 06111-1494

— — — — — — — — — — — — — — please fold and tape — — — — — — — — — — — — — — —